ENERGY SCIENCE, ENGINEERING AND TECHNOLOGY



# POWER ON: ELECTRIFICATION OF RURAL HEALTH, EDUCATION AND TOURISM FACILITIES

# ENERGY SCIENCE, ENGINEERING AND TECHNOLOGY

Additional books in this series can be found on Nova's website under the Series tab.

Additional E-books in this series can be found on Nova's website under the E-book tab.

ENERGY SCIENCE, ENGINEERING AND TECHNOLOGY

# POWER ON: ELECTRIFICATION OF RURAL HEALTH, EDUCATION AND TOURISM FACILITIES

WILLIAM R. THOMAS
AND
BENJAMIN E. WILSON
EDITORS

Nova Science Publishers, Inc.
*New York*

For permission to use material from this book please contact us:
Telephone 631-231-7269; Fax 631-231-8175
Web Site: http://www.novapublishers.com



**LIBRARY OF CONGRESS CATALOGING-IN-PUBLICATION DATA**

Power on : electrification of rural health, education, and tourism facilities / editors, William R. Thomas and Benjamin E. Wilson.
p. cm.
Includes bibliographical references and index.
ISBN 978-1-61209-202-7 (hardcover)
1. Rural electrification. 2. Small power production facilities. I. Thomas, William R. II. Wilson, Benjamin E.
TK4018.P648 2011
333.793'2--dc22

2010047629

*Published by Nova Science Publishers, Inc. † New York*

# CONTENTS

# PREFACE

Electricity is an increasingly essential commodity in remote healthcare, education and tourism facilities. Recent improvements in the distribution of vaccines and other cold chain-dependent supplies, as well as the global push to deliver anti-retroviral drugs and services to HIV-positive patients worldwide, have introduced new demands for electricity in sites with little or no access to reliable electrical power. This new book examines the challenging yet essential need to bring reliable energy to rural healthcare, education and tourism facilities across the globe.

Chapter 1- Electricity is an increasingly essential commodity in remote healthcare facilities. Recent improvements in the distribution of vaccines and other cold chain-dependant supplies, as well as the global push to deliver antiretroviral drugs and services to HIV-positive patients worldwide, have introduced new demands for electricity in sites with little or no access to reliable electrical power. Refrigerators and electronic diagnostic tools are part of the standard of care in many rural clinics throughout the world.

Chapter 2- In late June-early July 2009, the USAID Energy Team (EGAT/I&E/E) responded to a request by the USAID Senegal mission to assist its education program. USAID/Senegal is supporting the Government of Senegal (GOS) to construct new middle schools around the country, and some of the sites are in locations not currently connected to an electric grid. To date, 57 schools have been constructed under the PAEM program, and more schools are expected to be built in the future (under a different contractual framework). USAID/Senegal seeks a strategy on how to meet the energy needs of these schools, in order to ensure a quality education for students and inform future USAID programming in the education sector.

Chapter 3- Rural tourism is a rapidly growing sector of the global tourism industry. Since rural tourism operations are often in remote locations, ensuring a reliable supply of energy can be a special challenge. Therefore, planning for reliable energy is vital and impacts all decisions made at rural tourism facilities—since these areas may have difficult access to fossil fuel supplies and no access to grid electric power.

In: Power On
Editors: William R. Thomas et al.
ISBN: 978-1-61209-202-7


*Chapter 1*

# POWERING HEALTH: ELECTRIFICATION OPTIONS FOR RURAL HEALTH CENTERS*

***United States Agency for International Development, (USAID)***

> *"We work with a district hospital in Rwanda that often operates without any water or electricity. They have a huge generator but do not have the money to use it for a full day. They cannot run the lab equipment or the x-ray without starting the generator. The phone is powered by a small solar panel that has a bad battery so when there is cloud cover the phone disconnects. Lighting in the hospital is minimal."*
>
> – Jeroen van't Pad Bosch, Elizabeth Glaser Pediatric AIDS Foundation

## GOALS AND PRIORITIES: A BRIEF BACKGROUND

Electricity is an increasingly essential commodity in remote healthcare facilities. Recent improvements in the distribution of vaccines and other cold chain-dependant supplies, as well as the global push to deliver antiretroviral drugs and services to HIV-positive patients worldwide, have introduced new demands for electricity in sites with little or no access to reliable electrical

* This is an edited, reformatted and augmented version of United States Agency International Development, dated 2007.

power. Refrigerators and electronic diagnostic tools are part of the standard of care in many rural clinics throughout the world.

> *"Power disruptions result in a disruption of treatment programs or the inability to reliably use electrical appliances such as laboratory equipment. For those sites relying on generators, none of the sites reported being able to run the generators for long periods of time either due to the expense and difficulty of obtaining fuel or the lack of local mechanics to keep the motors running."*
>
> – Eddie Kariisa, Centers for Disease Control and Prevention, Rwanda

*Powering Health* is a resource for health professionals seeking to electrify health facilities that currently lack power or to ensure uninterrupted power for facilities that are connected to an unreliable power grid. Information is provided to help the user weigh the pros and cons of various energy systems with a focus on appropriate; solutions and special considerations for off-grid rural hospitals and health centers.

## *Powering Health* covers the following specific areas

- Gauging energy demand: tips on inventorying medical and other equipment to determine an estimate of a facility's energy needs.
- Understanding energy options: brief descriptions of solar, wind, reciprocating engine, and hybrid systems.
- Selecting an energy system: factors that influence the choice of a system, including facility size, level of care, budget, resource availability, and geographic location.
- Making investments sustainable: options for maintenance and financial planning to ensure the long-term sustainability and reliability of a facility's energy system.

Readers are encouraged to speak with experts to obtain the full range of information necessary to fulfill their energy needs in the most reliable and cost effective way. Representative cost figures are included for illustrative purposes, but actual prices may vary considerably in different regions. The basic maintenance needs for various technologies are discussed, but this guide is not a users' manual. In all cases, international development officials and clinic personnel must examine their facility's specific needs and discuss energy options with energy experts and local contractors.

# I. Reliable Energy: A Precondition to Success

Health professionals in rural clinics must address unexpected challenges on a daily basis. Poor infrastructure can delay the delivery of critical medical supplies, and inclement weather can make it difficult for patients to access the clinic. An unreliable energy source adds to these challenges. If the cold chain is inoperable when supplies arrive, vaccines, blood, and other medicines may go to waste. If a clinic is without lights, patients arriving at night must wait until morning to receive care. Selecting an appropriate source of reliable and sustainable energy can help mitigate some of the challenges inherent in operating a health facility in the developing world.

## Stepwise Approach to Electrifying Your Health Center

1. **Identify Your Health Center's Current Energy Demands**
   Fill out the blank chart in the back of the brochure with your specific information.
2. **Account for Near-Term Change**
   Determine whether your energy demands will change in the near-term.
3. **Maintaining and Financing Your Energy Technology**
   Institute financing mechanism(s) accounting for operating and maintenance needs and costs.
4. **Determine Technologies Needed to Meet Target**
   Evaluate energy technologies.
5. **Establish Target kWh/day**
   Fill out the blank chart in the back of the brochure with your specific information.
6. **Procure, Design System, and Install Technology**
   Select the most appropriate energy technology.

Remember to contact an expert for assessment, system design, procurement, installation, and maintenance of energy technologies.

# II. Understanding General Energy Needs

When considering the type of electrification needed to sustain daily operations, a facility must first understand its basic needs. The needs assessment will include an inventory of the types of equipment used in the facility and the power required to operate each device. Understanding the average "daily load", or the amount of power required to operate equipment under normal working conditions, will influence the choice of power supply. Once the daily energy requirement is established, a range of electrification options can be considered. Understanding the need will also provide managers with a realistic budget for procuring, installing, and maintaining the new system.

## Calculating Your Health Facility's Energy Needs

Appendices A and B can be used to help health professionals identify the overall energy demands of their facilities and match that demand with an appropriate electrification solution. Appendix A, *Identify Your Health Clinic Category*, provides energy consumption estimates for commonly used equipment in various types of health facilities. Appendix B, *Estimate Your Energy Use chart*, can be used to tailor the information in Appendix A to a specific facility. After filling in the chart with your specific data, you will gain an estimate of your facility's overall energy demand. This information, in addition to expert consultation, will assist in the selection of appropriate electrification technology.

*Strategic Thinking: Adjusting for Change*

Once a facility has comprehensively analyzed the energy requirements of its day-to-day operations, it must be determined whether those demands are likely to change. Facility managers must think strategically about the possibility that energy demands may increase due to the addition of patients, extended operating hours, or new services. Once this process has been completed and an adequate accounting of needs has been made, the manager can determine the various energy options to meet those demands. These options must be considered in light of all facility-specific variables.

## Categorization of Health Clinics

The following section describes several types of health facilities. The energy demands of a health facility will be a critical component in the selection of the most appropriate electrification technology. *Please note: these descriptions are provided as general comparative guidelines and are not precise descriptions of any one facility.*

### *Health Posts*

Health posts are the smallest, most basic health facility. These locations typically will not have a permanent doctor or nurse on staff. The health post may have a full- or part-time primary healthcare provider. Services available at health posts include the treatment of minor illnesses, the tending of minor injuries and, where possible, the provision of basic immunization services. Due to the limited medical equipment used, the overall energy demand of health posts is relatively low. The energy demands of a health post will be satisfied through *Category I Health Clinic* electrification options, while taking into account the reduced daily demand for energy.

### *Health Clinics*

Health clinics are generally larger than health posts and employ one or more full-time nurses. Clinics may also employ a part-time physician, depending on the size and location. A health clinic offers a wider array of services than a health post and will possess equipment allowing for more sophisticated diagnoses. Rural health clinics generally fall into one of three categories (*Categories I, II and III*), based on the type and number of medical devices used in the facility and the frequency with which they are used on a daily basis. Local resources may make specific energy options more or less advantageous in each location. The categories are listed on page five.

Other types of health facilities that require reliable and sustainable electrification include blood banks, stand-alone laboratories and pharmacies, and anti-retroviral treatment (ARV) clinics. Blood banks, stand-alone labs, and pharmacies will, depending on their size, utilize equipment similar to that found in *Category I or II Health Clinics* and will have similar energy needs. ARV clinics will have significant energy demands similar to those found in a *Category III Health Clinic* or higher. Energy requirements could be intensive for some ARV clinics due to the computer technology and additional equipment required to perform rapid blood analyses.

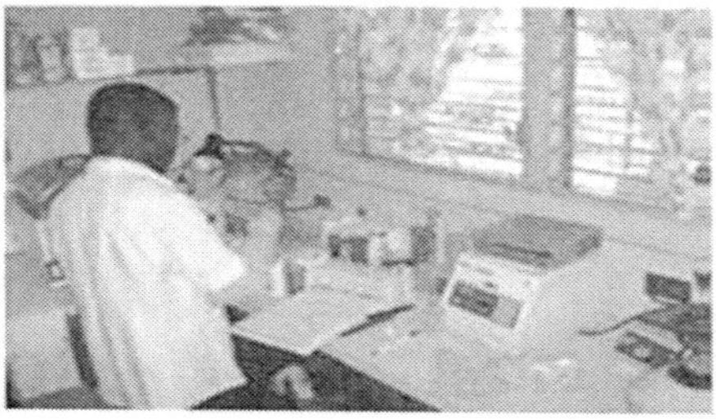

Lab technician preparing blood samples for grouping and cross-matching in a Haitian blood bank.

*Table I: Health Clinic Energy Needs* on page six illustrates the estimated cost of various energy technologies for a range of clinic sizes. In general, renewable energy options (e.g., photovoltaic (PV) system) will have higher capital costs than diesel or other fuel-based electricity generating options. However, over the long-term, renewable systems will have lower operating costs and produce fewer or no emissions. In renewable energy systems, battery maintenance, occasional cleaning, and theft-prevention will be the major recurring costs. A hybrid system using an alternative energy source (e.g., PV system) and a traditional generator (e.g., diesel) will have a higher up-front capital cost than a renewable-only system; however, hybrid systems provide greater flexibility, including the ability for one system to support the other. For illustrative purposes, a PV/diesel hybrid is represented in *Table I* on page six. Actual prices in a given location may vary considerably from those used in the table.

## Category I Health Clinic (Low Energy Requirements, 5 – 10 kWh/day)

- Typically located in a remote setting with limited services and a small staff
- Approximately 0 – 60 beds
- Electric power is required for:
  - lighting the facility during evening hours and to support limited surgical procedures (e.g. suturing)
  - maintaining the cold chain for vaccines, blood, and other medical supplies – one or two refrigerators may be used

- utilizing basic lab equipment – a centrifuge, hematology mixer, microscope, incubator, and hand-powered aspirator

## Category II Health Clinic (Moderate Energy Requirements, 10 – 20 kWh/day)

- Approximately 60 – 120 beds
- Medical equipment similar to Category I Health Clinic; frequency of use and number of devices are key factors of differentiation between Category I and II health clinics
- Separate refrigerators may be used for food storage and cold chain
- Communication device, such as a radio, may be utilized
- May accommodate more sophisticated diagnostic medical equipment and perform more complex surgical procedures

## Category III Health Clinic (High Energy Requirements, 20 – 30 kWh/day)

- Approximately 120 beds or more
- May serve as a regional referral center and coordinate communication between several smaller facilities and hospitals in large cities
- May need to communicate with remote health centers and hospitals by way of telephone, fax, computer, and Internet
- May contain sophisticated diagnostic devices (x-ray machine, CD4 counters, blood typing equipment, etc.) requiring additional power

Refer to Appendix A for a sample list of equipment and related energy requirements for the different categories of health clinics. Refrigeration, lighting, and computers are assumed to be energy-efficient; all other devices listed are standard (not energy-efficient). The equipment list and energy requirements represented in Appendix A are an approximation, designed to provide the reader with a short-hand framework to better understand the information presented below.

**Table I. Health Clinic Energy Needs.**

| 5kWh/DAY | | | | |
|---|---|---|---|---|
| **Technology** | **System Size** | **Capital ($)** | **Operating ($/year)** | **O&M Assumptions** |
| Solar Photovoltaic (PV) System with Batteries | 1200 W panels<br>20 kWh batteries | $12,000 system<br>$2,000 batteries | $500 | 1% of system cost per year (includes maintenance and component replacement, does not include security); Amortized cost of replacing the batteries every five years (20% of battery cost). |
| Wind Turbines with Batteries | 1,750 W turbine<br>20 kWh batteries | $10,000 system<br>$2,000 batteries | $600 | 2% of system cost per year; Amortized cost of replacing the batteries every five years. |
| Diesel Engine Generator | 2.5 kW | $2,000 | $1,400 | $0.0075/kWh maintenance, $0.67/kWh fuel ($1/liter for fuel is used), operating at 4kWh per day at 50% capacity, and replacement of engine every 10 years. |
| Hybrid Systems | 1,200 W panels<br>10 kWh batteries<br>500 W engine | $12,000 PV system<br>$1,000 batteries<br>$500 generator | $450 | 1% of PV system cost per year; battery replacement every five years; 200 hours of engine operation per year; replacement of engine every ten years. |
| Grid Extension | n/a | $10,000+ per mile | $200 | $0.10/kWh power. |
| **15kWh/DAY** | | | | |
| Solar Photovoltaic (PV) System with Batteries | 3,600 W panels<br>60 kWh batteries | $36,000 system<br>$6,000 batteries | $1,550 | Same as above. |
| Wind Turbines with Batteries | 5,250 W turbine<br>20 kWh batteries | $28,000 system<br>$6,000 batteries | $1,750 | Same as above. |
| Diesel Engine Generator | 2.5 kW | $2,000 | $3,900 | Same as above, operating at 15 kWh at 50% capacity. |

| 15kWh/DAY | | | | |
|---|---|---|---|---|
| **Technology** | **System Size** | **Capital ($)** | **Operating ($/year)** | **O&M Assumptions** |
| Hybrid Systems | 3,500 W panels 30 kWh batteries 1.5 kW engine | $35,000 PV system<br>$3,000 batteries<br>$1,000 generator | $1,350 | Same as above, with 200 hours of engine operation per year. |
| Grid Extension | n/a | $10,000+ per mile | $550 | Same as above. |
| **25kWh/DAY** | | | | |
| Solar Photovoltaic (PV) System with Batteries | 6,000 W panels 100 kWh batteries | $55,000 system<br>$10,000 batteries | $2,550 | Same as above. |
| Wind Turbines with Batteries | 8,750 W turbine 100 kWh batteries | $44,000 system<br>$10,000 batteries | $2,900 | Same as above. |
| Diesel Engine Generator | 2.5 kW | $2,000 | $6,400 | Same as above, operating at 15 kWh per day at 67% capacity. |
| Hybrid Systems | 6,000 W panels 50 kWh batteries 2.5 kW engine | $55,000 PV system<br>$5,000 batteries<br>$2,000 generator | $2,200 | Same as above, with 200 hours of engine operation per year. |
| Grid Extension | n/a | $10,000+ per mile | $900 | Same as above. |

# Energy Efficiency: Reducing Usage and Reducing Cost

As the table below indicates, a health facility can significantly reduce its energy demands by using energy-efficient appliances and devices. Typically, energy-efficient equipment and appliances are more expensive than standard-efficiency models. However, this higher cost is generally recouped through the reduced capital and operating costs of a smaller electricity generation system.

| Description | Power requirement or energy consumption of a high-efficiency model | Power requirement or energy consumption of a standard-efficiency model |
|---|---|---|
| Computer | 15 - 20 W (notebook computer) | 40 - 80 W (desktop computer without monitor) |
| Computer Monitor | 30 W (15" LCD monitor) | 65 - 120 W (15" to 21" CRT monitor) |
| Electric Lamp | 15 W (compact fluorescent lamp) | 60 W (incandescent lamp with a comparable light output) |
| Refrigerator/Freezer | 800 Wh/day | 1 ,800 to 2,500 Wh/day |

Installing more efficient equipment is an important component of energy conservation, but good management practices are equally important. These include maintaining equipment properly, insulating any areas that are heated or cooled, turning off unused lighting or equipment where possible, and monitoring energy consumption. All health center staff should be knowledgeable of the measures it takes to meet the center's energy needs and encouraged to help conserve energy.

Refrigerator storing blood and vaccines. Energy-efficient refrigerators can be used to conserve energy.

# III. Power Generation Options

After determining your facility's typical daily energy usage, it is time to evaluate the energy technologies available to electrify your facility. Rural health clinics have a number of options available to supply reliable electricity. The best option for a given application depends on a number of factors, and in some cases a combination of measures may be the best solution.

***Some factors to consider include***

- Reliability of local grid
- Local renewable energy resources (wind, solar, biomass)
- Local cost and availability of conventional energy resources (diesel, propane, gasoline)
- Local availability of systems, parts, service companies, and technicians
- Government policies and incentives
- System reliability requirements
- Technical capacity and funds for system maintenance and replacement
- Special considerations or desired operational characteristics – i.e., noise, emissions, etc.

***Options***

Key characteristics are discussed for the following technologies/options:

- Photovoltaic (PV)
- Wind
- Reciprocating Engines (generators)
- Hybrid Systems
- Grid extension

## Energy Technology Drivers

Capital Cost is the initial cost to purchase and install the equipment. In many cases, a portion of this cost may be met by grants or other forms of assistance. Clinics should be aware that power equipment – including generators, inverters, charge controllers, and batteries – can vary greatly in cost and quality. In many cases, higher-quality models will cost more, but can have a greater return on investment in terms of greater reliability of power and longer system lifetimes. Costs also vary considerably based on the local market. Research into specific brands is essential.

Operating Cost includes the cost of fuel (where applicable), operations and maintenance, security, and parts purchased for repairs. Costs of maintenance contracts and/or training of clinic staff should be factored into operating costs. Operating costs will vary more than capital costs, due to differences in:

- Fuel prices over time and from country-to-country.
- Use patterns. Systems will experience more or less stress in a given day, based on the number of hours they operate, the amount of power they provide, and the type of equipment drawing loads (e.g., high intensity equipment such as x-ray machines, as opposed to low-intensity equipment such as lighting).
- Environmental conditions.

**Durability** is the typical system lifetime, expressed either in years or (for engine generators) in hours of run-time.

**Reliability** is expressed as a fraction of time the equipment is available to provide power. Generators need to be taken off-line for service periodically, and wind and solar power systems require optimal weather conditions to operate at maximum efficiency. Systems can generally achieve greater reliability by adding backup components, although this generally increases cost and complexity.

**Emissions from generators include smog forming pollutants, particulate matter, and greenhouse gases.**

Special considerations exist for each technology. These may include resource availability, practical issues of deployment or operation, or risks and hazards associated with the equipment. Experts note that an important consideration common to many electrification programs is the cooperation of the host government at all levels. Expenses such as customs, permitting, or interconnection fees, as well as the associated delays with these processes, can make a system far more expensive than it would be otherwise.

**Photovoltaic (PV) Systems** generate electricity from sunlight collected by solar panels. Energy collected in this manner can be used to supply direct power to electrical equipment, or it can be stored in batteries to provide indirect power. In combination with a battery, PV systems are excellent for handling small-to-moderately sized loads in locations where grid electricity is not available. They are highly modular, so it is easy to customize the system to your needs and add units if your power demand increases. PV systems are quiet and produce no emissions. Photovoltaic systems provide DC (direct current) power. Inverters, which convert this DC power to AC power, must be added to most systems powering medical equipment used in Category II or higher health clinics.

PV systems are rated in terms of the peak power (in watts or kilowatts) they can produce. This is multiplied by the peak sun hours (PSH) factor to determine the energy (in kilowatt-hours) produced each day. An optimal setting in Africa might receive the equivalent of five or more hours of peak sunlight per day. In this case, a 1-kW system would produce 5 kWh per day. There is always some loss of power from rated output, so in our analysis we have conservatively increased system sizes by 20% over the rated load to ensure power supplies are sufficient to meet the identified needs.

### *Cost Considerations*

Depending on the size of the system, installed PV systems can cost $8.00-12.00/ watt. However, unusual expenses such as transportation of modules, customs fees, or permitting expenses can increase this cost. If the system has no other backup, the battery must be sized appropriately to provide power after several days of overcast weather; therefore, local weather patterns can influence overall system cost. In a hybrid configuration, the batteries can be much smaller since the generator can take over if there are prolonged cloudy periods. PV panels are the most expensive part of a solar electric system; as such, they are sometimes targets for theft. Vandalism can also be a problem. For larger loads, their high capital cost can render them a less preferable option if grid extensions or fuel for generators are readily available.

> *"Based on 121 health centers we visited in Rwanda, 49% utilized solar energy. Most of the existing water and electricity equipment is functional, with the exception of solar energy equipment; in approximately one-third (37%) of health centers, the solar equipment was not functional at the time of the study due to forgone maintenance."*
>
> -Laura Hoemeke, IntraHealth International

Solar Photovoltaic (PV) System.

***Maintenance Considerations***

PV systems have been the focus of numerous efforts for rural electrification. The panels themselves typically have a very long lifetime (20-30 years). Unfortunately, installation programs do not always include a sufficient service component. Many health centers describe experiences with PV systems with inoperative batteries, resulting in, for example, phones that only operate when the sun is shining. Regular maintenance on batteries is essential; they should be checked every month, with the electrolyte level replenished as needed. Properly maintained, batteries should last several years before needing replacement. While training local hospital staff in system maintenance is essential for routine maintenance, a professional technician should also perform an annual maintenance check, examining wiring connections, mounting bolts, and inverter operation.

## PV System Lessons Learned

- PV systems typically have higher capital costs, but lower operating costs when compared to other energy generation options.
- The availability of replacement components (model and brand) from local vendors should be considered when procuring initial system components.
- End-user expectations of solar systems are often unrealistic – education on the practical application of solar systems must accompany system design and installation.
- National standards for the placement, design, procurement, installation, and servicing of photovoltaic systems can help improve sustainability.

- Donor-funded PV systems often fail for lack of operating funds and local service infrastructure.
- Detailed user manuals are critical – especially in cases where staff turnover is high.

Local ownership, often established through a contribution to initial system cost, is critical for system sustainability.

## DETERMINING LOCAL RESOURCE AVAILABILITY

Renewable resource availability is a key consideration when choosing an appropriate energy generation technology. For additional information regarding local resource availability, please see the following resources:

**Solar Insolation**

The National Aeronautics and Space Administration (NASA) provides extensive information on solar resources by month for any location in the world. This data includes average sunlight (in kWh per square meter per day), typical number of consecutive sunless days, and the minimum sunlight typically available each month. This resource, or one of similar quality, should be consulted when designing a system including PV. http://eosweb.larc.nasa.gov/sse/

**Wind Resources**

Locally, wind resources will vary more than solar resources. If no other wind systems are located nearby, then a contractor should conduct monitoring at the site for at least several months prior to installation. http://www.wwindea.org/home/index.php

http://www.afriwea.org/

**Hybrid Systems**

The HOMER tool, developed by the U.S. National Renewable Energy Laboratory (NREL), is useful for designing and modeling hybrid systems. The HOMER tool automatically retrieves solar insolation information from NASA. http://www.nrel.gov/homer/

## Table II. Energy Technology Characteristics.

Refer to the table below for an illustration of key characteristics of energy generation technologies. Capital cost, operating cost, reliability, emissions, resource availability, and other factors should be considered when selecting an energy technology.

| Energy Technologies | | Capital Cost | O&M Cost | Reliability | Durability | Special Considerations | Emissions | Optimal Use |
|---|---|---|---|---|---|---|---|---|
| Solar Photovoltaic (PV) System with Batteries | | Very high | Low | High (if maintained properly) or low (if not) | 20 - 30 years (PV) 5 years (batteries) | Theft (batteries or panels); Vandalism (panels); Availability of trained technicians | None | Small loads; Areas where fuel is costly or difficult to obtain |
| Wind Turbine with Batteries | | High | Low-moderate | High (if maintained properly) or low (if not) | 20 years (turbine); 10 years (blades) 5 years (batteries) | Theft (batteries); Lack of data on wind resources | None | Many moderate loads where resource is sufficient |
| Reciprocating Engines | Diesel Generator | Moderate-high | High | High | 25,000 operating hours | Fuel spills; Emissions | Very high | Larger loads |
| | Gasoline Generator | Low | Very high | Moderate | 1,000 - 2,000 operating hours | Fuel spills; Emissions; Flammability | High | Emergency generator |
| | Gas Generator | Moderate | High | Moderate | 3,000 operating hours | Propane is of limited availability, but can use biogas | Low | Component in hybrid system or stand-alone |
| **Hybrid System** | | Very high | Low-moderate | Very high | Varies; Optimization greatly extends generator and battery life | Complexity for servicing | Low | Medium and larger loads |
| **Grid Extension** | | Varies | None | Varies | High | Theft; Extending grid allows connection of nearby homes to grid | Not local | Where grid is reliable and not too distant |

**Wind Turbines** generate emissions-free electricity from a renewable energy resource. Situating a wind turbine on a site requires careful measurement of wind patterns. While extensive maps of solar resources exist, the same is not true of wind resources. Small wind turbines are available with capacities in the range of a few hundred watts. Larger turbines offer significant economies of scale. Similar to PV systems, wind turbines must be connected to the grid or to a battery system to provide reliable power.

***Cost Considerations***

Installed wind systems typically cost $4.00-6.00/watt for small systems, with the cost per watt decreasing substantially as turbine size increases. The capital cost for wind turbines is lower than PV systems for moderate-to-large loads. Like PV, wind systems do not require fuel.

***Maintenance Considerations***

Maintenance requirements for turbines are heavily dependent on local conditions. Due to the moving parts, maintenance for wind turbines is somewhat higher than for PV systems. Normally, a turbine should receive a maintenance check every three months. Necessary maintenance includes lubrication of the moving parts, checking bolts and electrical connections, checking supporting guy wires for proper tension, and examining components for corrosion. Rotor blades and bearings normally need replacement after about ten years. Small turbines (fewer than 10 kW) generally do not require a gearbox, which is a component that often requires attention in larger systems.

Wind Turbine.

Reciprocating Engine (Generator).

**Reciprocating Engines** (small generators) are the most common form of small generator. A reciprocating engine is similar to an automobile engine, using the combustion of a fuel to drive pistons. In generator-only systems, the generator must be sized to handle the peak expected load, but the system frequently runs at lower loads at reduced efficiency.

***Cost Considerations***

Engine generators typically have low capital cost compared to other alternatives, but higher operating costs due to the need for fuel. Initial capital cost will be approximately $2,000 for 2.5 kW, varying with the size and type of generator. Operating costs will vary depending on the level of usage. Engines do offer economies of scale in increasing sizes, but utilizing a few smaller units is generally preferred. This allows cycling of the generators for maintenance, and allows the use of one or two units at full load rather than a larger unit at reduced load. A battery bank can also be added to a generator-based system to reduce run time and save fuel costs.

## FUEL OPTIONS FOR GENERATORS

**Diesel** is the most common option for generator fuel. Diesel engines tend to be more expensive than gasoline generators, but also more reliable and longer-lived. The fuel does not burn as cleanly as propane or biogas.

**Gasoline** is widely available and is used in many generators. Gasoline generators are cheaper than diesel generators and available in smaller sizes. They are normally used as emergency back-up generators due to their shorter operating lifetimes.

**Propane,** where available, is another generator fuel. Propane generators are quieter, cleaner, and environmentally safe; spilled propane evaporates rather than contaminating a site. Propane generators are well-suited to use in a hybrid system with solar or wind, though they are not optimal for serving as the sole (or primary) energy source. Propane may also be used to run a refrigerator directly. In this application, the refrigerator is designed to run on the heat of propane combustion rather than on electricity. Use of a propane generator as part of a hybrid system, and using electricity to run the refrigerator, allows greater versatility and a wide range of appliances to be powered by the system.

***Maintenance Considerations***

Engines have several maintenance requirements. Engine oil and the oil filter should be changed after approximately 1,000 hours of operation. The engine head may need to be rebuilt after 8,000 hours, and the engine block after 16,000 hours. Facility personnel can examine the generator daily for fuel, oil, or coolant leaks. The area in the vicinity of the generator should be kept free of any debris that could pose a fire hazard or make access difficult. Oil and coolant levels should be checked weekly. More in-depth monthly, semi-annual, and annual maintenance checks should be performed by professionals. These maintenance checks will vary depending on the design of the engine and its duty cycle (whether it is a primary energy source or a back-up/emergency unit).

The expertise to service diesel and gasoline generators is widespread because the units are similar, but not identical to, automotive engines. In some cases, spare parts for generators may be difficult to obtain.

## Reciprocating Engine Lessons Learned

- Reciprocating engines typically have lower capital costs, but higher operating costs when compared to renewable technologies.
- Fuel availability and cost are key considerations for this technology.

**Hybrid Systems** employ a number of different technologies. A system could include photovoltaic panels, a wind turbine, batteries, and a generator. With reliable solar and wind resources, this system would rarely need to rely on a generator. Since the generator can recharge the batteries during prolonged periods of inclement weather, the battery bank in a hybrid system can be significantly smaller than a PV-battery system, perhaps only needing to store one or two days' worth of energy. The low duty cycle extends generator lifetime.

### *Cost Considerations*

The capital cost of a hybrid system is much higher than a system using only propane generators (or diesel generators); however, the hybrid system shows significant savings over a 20- year lifetime, primarily due to savings on maintenance and fuel costs. The lifetime of a generator in a hybrid system is extended because it does not have to run as often. When emission savings are considered, the benefits of the hybrid system become even more apparent. For a more detailed explanation, refer to Case Study I: Evaluating Energy Generation Options.

### *Maintenance Considerations*

A monthly check is recommended for hybrid systems. The generator is not designed for heavy use within a hybrid system, but still requires periodic attention. A regular maintenance program will also help maintain personnel familiarity with the system's proper operation.

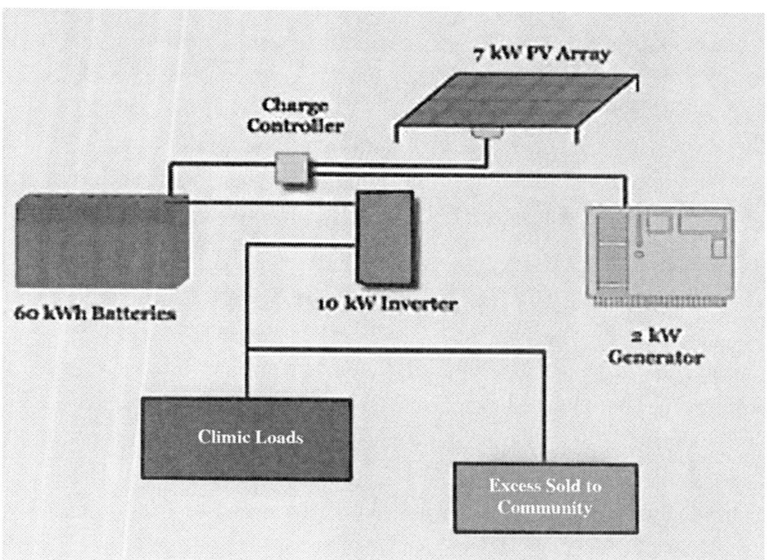


Figure I. Hybrid System.

**HYBRID SYSTEM LESSONS LEARNED**

- Hybrid systems (including renewable energy, batteries, and back-up generators) are a feasible option for many medium-to-large loads.
- Reduced operating costs will offset high initial capital costs over time.
- The increased complexity of hybrid systems requires sound maintenance programs and trained technicians.

**Grid Extensions** connect the facility to the local utility. This involves running power lines across what is, for many rural clinics, fairly difficult terrain. Grid extension varies in cost depending on the utility, the terrain, the distance to be covered, and the size of the load to be served. If the local grid is fairly reliable and within a few miles, obtaining an estimate for the capital cost of grid extension and the recurring cost of electricity will give points of comparison when considering other options. However, if grid electricity is less than reliable, then some form of backup energy will be required even with a grid extension.

### *Cost Considerations*

Grid extension typically has a high capital cost, but offers several advantages. First, in most cases, the cost varies only slightly with the size of the load to be served. This makes it uneconomical for small loads but a better option for large loads. Second, once the grid has been extended to the clinic, it can serve other local loads. Third, operations and maintenance are the responsibility of the utility.

Grid Extension.

# Key Components of PV, Wind, and Other Systems

## Battery

Batteries are not a power technology, but a means of storing the power produced by other systems. They are frequently used in conjunction with photovoltaic or wind systems, and may be used in other systems as well. Lead-acid batteries are widely available and relatively inexpensive; however, they have shorter lifetimes than newer alternatives. Lead-acid batteries need the electrolyte replenished with distilled water periodically (tap water may contain minerals that harm battery performance and lifetime). Electrolyte levels should be checked every month, and new water is normally needed approximately every three months, though this varies with local conditions.

For some grid-connected clinics, batteries can be used to provide backup power during outages. However, if grid power is absent for a lengthy period of time, some other system would be useful to recharge the batteries. Batteries' lifetimes are partly dependent on the cycling (charging and discharging) they experience, but even batteries that are not heavily used will deteriorate over time.

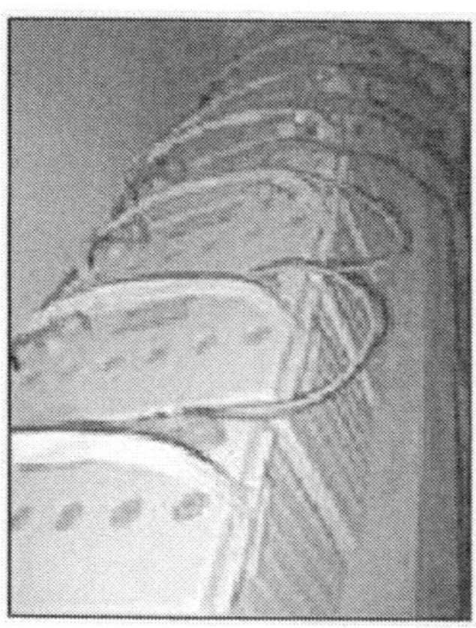

Battery banks provide additional power during surges and outages.

Battery maintenance is fairly simple and not overly time-consuming, but it is critical. The battery plates must remain submerged in the acid and cannot become exposed to the air or severe damage to the system can result. Since only the water evaporates out of the battery, only water needs to be replaced (unless there is a spill). Health centers should budget for the expense and travel associated with acquiring distilled water. Excessive heat can also shorten battery lifetime. Typical battery life is approximately 5 years.

Disposal of batteries at the end of system life is an important consideration. Lead-acid batteries can contaminate groundwater supplies if improperly disposed. Recycling of batteries at an environmentally sound facility is a preferable option.

### Inverter

An inverter is a component of any system involving PV or batteries and is used to convert direct current (DC) to alternating current (AC) which is often required by certain medical equipment. Inverters have different maintenance needs depending on their design. They should be examined for proper operation at least every three months, if not more often. Similar to batteries, inverters are best kept at a moderate temperature. In many cases, replacement parts will not be locally available and will have to be ordered from manufacturers.

### Charge Controller

A charge controller regulates the voltage and current coming from the solar panels (or wind or generator) to the battery to prevent overcharging or over discharging. Most modern controllers maintain system voltage regulation electronically by varying the width of DC pulses they send to the batteries. The wider the pulse, the more power goes to the batteries.

## IV. System Sustainability

## Importance of Maintenance

- Regular and timely maintenance of all electrification equipment is essential to proper functioning of the equipment.
- Routine maintenance, as well as major overhauls and capital

replacement, need to be planned and budgeted for in advance.

- Lack of maintenance ultimately will have a negative impact on reliability of power supply.
- Maintenance problems often are easily preventable, yet frequently overlooked.
- Emergency back-up generators should be checked periodically even if rarely used.
- Improper or insufficient maintenance can lead to substantial costs in the future.

Regular maintenance is well worth the cost, and programs installing energy systems at health clinics or other facilities should ensure there will be a commitment to servicing the system. Experts recommend training local personnel in the servicing of these systems or obtaining a long-term maintenance contract.

While proper installation is important, ongoing maintenance is also a key factor.

## Financing: Innovative Options

Clinic managers must develop a sustainable way to pay for the maintenance and operation of the system to ensure continuity of facility operations. Facilities should consider incorporating aspects of the innovative finance structures described below into their financial and operating practices.

### *User Fees*

A "user fee" system involves building the cost of energy into the overall cost of medical services – passing the cost to the patient. Most rural medical clinics struggle to secure sufficient operating funds due to the inability to pass along true costs of medical service to users who lack the resources to pay actual costs. The inability of patients to pay, coupled with the challenge of managing the collection and disbursement of funds, makes this approach difficult to implement.

### *Sale of Excess Electricity*

The sale of excess electricity offers a promising approach to finance operations. By installing a system with excess capacity, income from the sale of additional power can offset a portion, if not all, of the system's operating costs.

### **Bulk (Wholesale)**

By operating the power system as a small enterprise, excess electricity can be sold to nearby villages, factories, schools or facilities. The system must be sized to accommodate both the clinic and the potential customer base. Customers must be in close proximity to the system or transmission costs quickly make this approach prohibitively expensive. Maintenance requirements are also more complex.

> *"The Center de Sainte Bushara Health Clinic in Rwanda serves a population of about 21,000 people and treats an average of 1,300 patients each month. Its 14 PV systems are used to provide lighting, refrigeration, and power for some diagnostic equipment for the facility. The hospital is funded largely by a community health cooperative. Each year, local residents pay 3,500 Rwanda francs into the coop, and are then treated for free if they need medical services. A portion of the money is being budgeted for maintenance and upkeep of the PV systems."*
>
> Source: Solar Light for Africa, 2006

### **Point of Use Sale (Retail)**

When potential purchasers of power are too remote to obtain the electricity over transmission lines, the clinic can sell it at or near their facility. A small powering station can be established with fees charged based on the amount of power used, if metering is available, or according to time. For customers with transportable devices such as power tools, a small work area with outlets can be set aside adjacent to the power station where users can plug in equipment. Villagers can use these areas for income-generating activities.

It may be feasible to establish a "mini industrial zone" near the clinic's power system, providing an area with permanent workshops (sewing, weaving or repair services) or stores. The clinic could realize income from rent on the workshop/store space, the sale of electricity, and the pumping of water.

***Institutional Management***

Establishing an entity that has a stake in the continued successful operation of the system is crucial to cultivating a sense of ownership for on-going system operation. Innovative financing systems must be properly managed by organizations and individuals who use and pay for the power.

Management structures include existing clinic management, nearby villages or facilities, or a new organization dedicated to providing oversight of the energy system such as a cooperative between villages. The cooperative can include an agreement with the clinic to manage any of the financing arrangements previously described. The level of responsibility of the cooperative can range from total operation and management of the system to simply keeping track of usage and payments.

## WRAPPING UP

This publication has introduced fundamental issues to be considered when addressing the energy needs of rural, off-grid health facilities. Health professionals investigating options for improving energy services at health facilities should keep in mind:

- Many electrification options are available – each has advantages and disadvantages that must be carefully considered within a specific context.
- Energy generation systems should be designed to meet the specific needs of the health facility.
- Purchase of inappropriate technologies, incorrectly sized "off-the-shelf" systems, and insufficient attention to maintenance programs can be costly and result in system failure.
- Hybrid systems (including renewable energy, batteries, and fossil-fueled generators) are a feasible option for many medium-to-large

loads.

- The sale of excess electricity (or pumping clean water) offers a promising approach for health facilities to finance operations and maintenance costs.

# APPENDIX A. IDENTIFY YOUR HEALTH CLINIC CATEGORY.

| **Category I Health Clinic Quantity Power and Energy Consumption** | **A** | **B** | **C = AxB** | **D** | **E = CxD** | **F = E/1000** |
|---|---|---|---|---|---|---|
| | **Quantity** | **Power (Watts)** | **Total Watts** | **On-Time (watt hours/day)** | **Wattage (hours/ day)** | **kWh/day** |
| | **1** | **60** | **60** | **6.0 - 12.0** | **360 - 720** | **0.36 - 0.72** |
| Small Refrigerator (non-medical use) | 1 | 300 | 300 | 5.0 | 1,500 | 1.5 |
| Centrifuge | 1 | 575 | 575 | 1.0 - 1.5 | 575 - 862.5 | 0.575 - 0.862 5 |
| Hematology Mixer | 1 | 28 | 28 | 1.0 - 1.5 | 28 - 42 | 0.028 - 0.042 |
| Microscope | 1 - 2 | 15 | 15 - 30 | 3.0 - 4.0 | 45 - 120 | 0.045 - 0.12 |
| Lighting | 2 | 10 | 20 | 2 - 10 | 40 - 200 | 0.040 - 0.2 |
| Incubator | 1 | 400 | 400 | 2.0 - 12.0 | 800 - 4,800 | 0.8 - 4.8 |
| Water Bath | 1 | 1,000 | 1,000 | 1.0 - 2.0 | 1,000 - 2,000 | 1 - 2 |
| Communication via VHF Radio | 1 | | | | | |
| Stand-by | | 2 | 2 | 12.0 | 24 | 0.024 |
| Transmitting | | 30 | 30 | 1.0 - 2.0 | 30 - 60 | 0.030 - 0.060 |
| **Total** | | | **2,430 - 2445** | | **4,402 - 10,328.5** | **4.4 - 10.3** |

## Appendix A. (Continued).

| Category II Health Clinic Quantity Power and Energy Consumption | A | B | C = AxB | D | E = CxD | F = E/1000 |
|---|---|---|---|---|---|---|
| | Quantity | Power (Watts) | Total Watts | On-Time (watt hours/day) | Wattage (hours/ day) | kWh/day |
| | **1** | **60** | **60** | **6.0 - 12.0** | **360 - 720** | **0.36 - 0.72** |
| Vaccine Refrigerator/Freezer | 2 | 60 | 120 | 6.0 - 12.0 | 720 - 1,440 | 0.72 - 1.44 |
| Small Refrigerator (non-medical use) | 1 | 300 | 300 | 5.0 | 1,500 | 1.5 |
| Centrifuge | 1 | 575 | 575 | 2.0 | 1,150 | 1.15 |
| Hematology Mixer | 1 | 28 | 28 | 2.0 | 56 | 0.056 |
| Microscope | 2 | 15 | 30 | 5.0 | 150 | 0.15 |
| Lighting | 2 - 3 | 15 | 30 - 45 | 10.0 | 300 - 450 | 0.3 - 0.45 |
| Sterilizer Oven (Lab-oratory Autoclave) | 1 | 1,564 | 1,564 | 1.0 | 1,564 | 1.564 |
| Incubator | 2 | 400 | 800 | 2.0 - 12.0 | 1,600 - 9,600 | 1.6 - 9.6 |
| Water Bath | 1 | 1,000 | 1,000 | 2.0 | 2,000 | 2.0 |
| Communication via VHF Radio | 1 | | | | | |
| Stand-by | | 2 | 2 | 12.0 | 24 | 0.024 |
| Transmitting | | 30 | 30 | 2.0 - 3.0 | 60 - 90 | 0.060 - 0.09 |
| Desktop Computer | 1 - 3 | | | | | |
| Power Save Mode | | 50 | 50 - 150 | 7.0 | 350 - 1,050 | 0.35 - 1.05 |
| In Use | | 150 | 150 - 450 | 5.0 | 750 - 2,250 | 0.750 - 2.25 |
| Printer | 1 | 65 | 65 | 3.0 | 195 | 0.195 |
| **Total** | | | **4,744 - 5,159** | | **10,569 - 21,519** | **10.6 - 21.5** |

## Appendix A. (Continued).

| Category III Health Clinic Quantity Power and Energy Consumption | A | B | C = AxB | D | E = CxD | F = E/1000 |
|---|---|---|---|---|---|---|
| | Quantity | Power (Watts) | Total Watts | On-Time (watt hours/ day) | Wattage (hours/ day) | kWh/day |
| Vaccine Refrigerator/Freezer | 3 | 60 | 180 | 6.0 - 12.0 | 1,080 - 2,160 | 1.08 - 2.16 |
| Small Refrigerator (non-medical use) | 1 | 300 | 300 | 5.0 | 1,500 | 1.5 |
| Centrifuge | 1 | 575 | 575 | 3.0 | 1,725 | 1.725 |
| Hematology Mixer | 1 | 28 | 28 | 3.0 | 84 | 0.084 |
| Microscope | 3 | 15 | 45 | 6.0 | 270 | 0.27 |
| Lighting | 3 - 4 | 15 | 45 - 60 | 10.0 | 450 - 600 | 0.45 - 0.6 |
| Incubator | 3 | 400 | 1,200 | 2.0 - 12.0 | 2,400 - 14,400 | 2.4 - 14.4 |
| Water Bath | 1 | 1,000 | 1,000 | 2.0 | 2,000 | 2.0 |
| Communication via VHF Radio | 1 | | | | | |
| Stand-by | | 2 | 2 | 12.0 | 24 | 0.024 |
| Transmitting | | 30 | 30 | 3.0 - 4.0 | 90 - 120 | 0.090 - 0.120 |
| Desktop Computer | 3 - 6 | | | | | |
| Power Save Mode | | 50 | 150 - 300 | 7.0 | 1,050 - 2,100 | 1.05 - 2.1 |
| In Use | | 150 | 450 - 900 | 5.0 | 2,250 - 4,500 | 2.25 - 4.5 |
| Printer | 1 | 65 | 65 | 4.0 | 260 | 0.26 |
| Portable X-Ray Machine | 1 | 3,000 | 3,000 | 1.0 | 3,000 | 3.0 |
| **Total** | | | **8,634 - 9,249** | | **17,597 - 34,157** | **17.6 - 34.2** |

## APPENDIX B: ESTIMATE YOUR ENERGY USE.

| **Description of Device** | **A Quantity** | **B Power[1] (Watts)** | **C = AxB Total Watts** | **D On-Time[2] (hours/day)** | **E = CxD Wattage (hours/day) (total watts x on-time)** | **F = E/1000 kWh/day** |
|---|---|---|---|---|---|---|
| | | | | | | |
| | | | | | | |
| | | | | | | |
| | | | | | | |
| | | | | | | |
| | | | | | | |
| | | | | | | |
| | | | | | | |
| | | | | | | |
| | | | | | | |
| | | | | | | |
| **Other Factors to Consider** | **Comments** | | | | | |
| Number of Beds | | | | | | |
| Average Patients Served/Day | | | | | | |
| Access to Telecommunication | | | | | | |
| Solar Resource[3] | | | | | | |
| Wind Resource[4] | | | | | | |
| Road Quality | | | | | | |
| Availability and Cost of Gasoline | | | | | | |
| Availability and Cost of Propane | | | | | | |
| Availability and Cost of Diesel | | | | | | |

[1]To find wattage information for a given device, check the user's manual or it may be written on the back of the device.

[2]The typical daily "on-time" for a standard refrigerator compressor is 5-6 hr/day.

[3]Refer to the following for solar resource data: http://eosweb.larc.nasa.gov/sse/

[4]Refer to the following for wind resource data: http://www.wwindea. org/home/index.php.

Whether designing a national health facility retrofitting program or choosing the appropriate technology for one specific site, readers are strongly encouraged to seek independent expert advice. This expert advice can help ensure that a system is appropriately designed and that proper safeguards are in place to ensure sustainability.

## GLOSSARY

**Alternating Current (AC)** – An electric current in which the direction of flow oscillates at frequent, regular intervals.

**Amp** – A measure of electrical current flow.

**Battery** – A device that stores energy and makes it available in an electrical form.

**Capital Cost** – The initial cost to purchase and install equipment.

**Charge Controller** – Controls the flow of current to and from the battery to protect from over charging or over discharging.

**Cold Chain** – System of people and equipment that attempts to keep vaccines and blood at proper temperatures as they are distributed from the manufacturer or supplier to the locations where they are administered.

**Direct Current (DC)** – An electric current flowing in one direction.

**Durability** – The typical system lifetime, expressed either in years or (for engine generators) in hours of run-time.

**Electricity** – Energy made available by the flow of electric charge through a conductor.

**Energy** – The capacity of a physical system to do work. The units of energy are joules.

**Grid** – The network of transmission lines, distribution lines, and transformers used in central power systems.

**Insolation** – The amount of sunlight falling on an area over the course of a year, often measured in watts per square meter.

**Inverter** – A solid state device that produces an AC output from a DC input.

**Kilowatt (kW)** – One thousand watts.

**Kilowatt Hour (kWh)** – The work performed by one kilowatt of electric power in one hour.

**Load** – The amount of electric power or energy delivered or required at any specified point or points on a system.

**Operating Cost** – The day-to-day expense of using and maintaining property.

**Photovoltaic (PV) System** – The production of electricity from sunlight, commonly referred to as "solar electric".

**Power** – The rate of doing work; measured in watts.

**Renewable Energy** – Energy derived from non-fossil fuel resources; includes energy produced from PV, wind turbines, hydroelectric, and biomass.

**Solar Electric** – See photovoltaic system.

**Volt, Voltage (V)** – A unit of electrical force or electric pressure.

**Watt, Wattage (W)** – A unit of power equal toone joule per second.

**Watts** = volts x amps.

**Wind Turbine** – A device that converts the energy of moving air into electricity.

**Yield** – Output: the quantity of something (as a commodity) that is created (usually within a given period of time).

Note: The following two case studies collectively illustrate the stepwise approach to electrification of health facilities that has been discussed in this guidebook. In the first example, a hypothetical clinic in Botswana is used to demonstrate the utility of a modeling tool to assist in the critical task of system design. The second case study details the electrification of a hospital in Uganda utilizing a PV system to meet the energy needs of the clinic and provide clean drinking water to the community. Strategies to ensure proper operation and maintenance of these systems are also highlighted.

# CASE STUDY I: EVALUATING ENERGY GENERATION OPTIONS

## A Hypothetical Case Study for a Rural Health Clinic in Botswana

A hypothetical clinic in the Kalahari Desert of Botswana currently has a small refrigerator, lights, hematology mixer, microscope, computer, and communications equipment. Using the worksheet in Appendix B it is determined that the average daily load of the clinic is 13 kWh per day with additional load of 2 kWh per day expected in the near future. The clinic is not connected to the grid and currently utilizes a diesel generator to partially meet its energy needs. An international donor agency has been working with this clinic to improve local health care service delivery and would like to explore different options for upgrading its power generation systems.

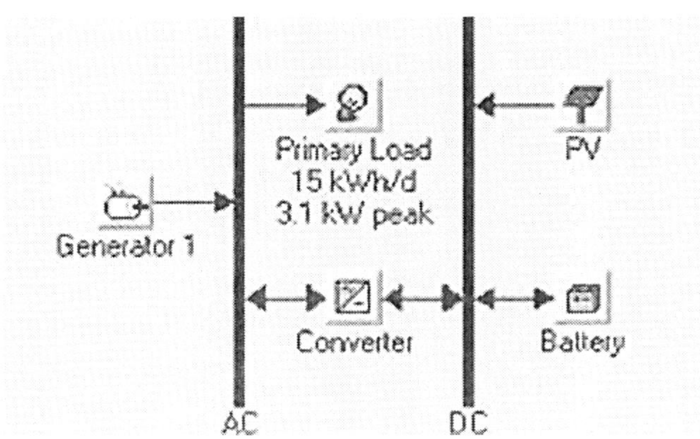

Several different modeling tools have been developed which allow a user to compare different energy generation options for this facility. For this example, we use the HOMER program, specifically developed by the U.S. National Renewable Energy Laboratory to analyze stand-alone systems that include renewable energy components.[5]

The HOMER model can be used to compare costs for a variety of different energy generation systems that can meet 100% of this clinic's load. We considered systems with combinations of the components shown in the figure above: generator, PV, AC/DC converter (inverter) and batteries.

In addition to the load data already calculated for this clinic, other site-specific information needed for this model is renewable resource availability, cost of fuel, and component costs. Appropriate solar radiation data for this location is automatically accessed by HOMER from the NASA database at the web site http://eosweb.larc.nasa.gov/sse/. Component costs can be estimated based on in-country information.

The resulting cost estimates are shown in the table below, which ranks a variety of alternative system designs by the lifetime cost of energy per kWh. The lowest cost system is a PV-dieselbattery hybrid system. The calculations demonstrate that because of fuel and maintenance costs, the system with the lowest capital cost is not the system with the lowest lifetime cost of energy. A diesel-battery system costs 13% more than this hybrid system because the added fuel cost over the life of the system is more than the savings in initial PV investment, and a PV-battery system costs about 28% more than the least-cost design. Notice that the cost of energy from a diesel system with no batteries is over twice the cost from a diesel- battery system. The addition of batteries to a diesel system are often a good investment in terms of fuel savings.

**Comparison of Alternative System Designs, diesel price $0.80/L.**

| Components | PV (kW) | Diesel (kW) | Battery (2 kWh) | Converter (kW) | Initial Capital | Total NPC | Lifetime (25 years) Cost of Energy ($/kWh) |
|---|---|---|---|---|---|---|---|
| PV, Diesel, Battery | 3 | 1 | 12 | 2 | $35,050 | $60,957 | 0.67 |
| Diesel, Battery | - | 2 | 16 | 1.5 | $6,000 | $69,008 | 0.759 |
| PV, Battery | 4 | - | 30 | 3 | $49,000 | $78,177 | 0.859 |
| Diesel | - | 3 | - | - | $1,950 | $153,946 | 1.692 |

The total net present value of costs (NPC) of the least-cost system and its components is shown below. The fuel cost component is very small because the generator runs only 113 hours during the year. Notice that all technologies have significant O&M requirements, totaling about $300 per year. If sufficient funds are not available to cover these maintenance costs, or if trained technicians are not accessible, the systems will not be sustainable.

| Component | Initial Capital ($) | Annualized Capital ($/yr) | Annualized Replacement ($/yr) | Annual O&M ($/yr) | Annual Fuel ($/yr) | Total Annualized ($/yr) |
|---|---|---|---|---|---|---|
| PV Array | 30,000 | 1,805 | 689 | 75 | 0 | 2,569 |
| Generator | 650 | 39 | 17 | 146 | 186 | 388 |
| Batteries (12) | 2,400 | 144 | 298 | 60 | 0 | 502 |
| Converter | 2,000 | 120 | 86 | 3 | 0 | 209 |
| Totals | 33,050 | 2,109 | 1,089 | 284 | 186 | 3,668 |

As discussed in this guidebook, the lifetime cost of the system is one of many important considerations when choosing the appropriate technology to power a health clinic. When considering the pros and cons of different system designs, a modeling program such as HOMER is a valuable tool.

## CASE STUDY II: KALUNGI HOSPITAL WATER AND ELECTRIFICATION PROJECT

Location: Kalungi, Uganda, 2006

Partners: The Coca-Cola Company/Solar Light for Africa, Ltd. (SLA)/Solar Energy Uganda, Ltd. (SEU)/ Geneva Global Foundation/Global Environment & Technology Foundation (GETF)/United States Agency for International Development

Total Electrification and Water Project Cost: $121,000 (Includes contractor and equipment costs, excludes overhead/administrative costs)

Source: Solar Light for Africa, 2006

### *Overview*

Kalungi Hospital is located 125 kilometers south of Kampala, Uganda, off the Kampala-Masaka Highway. The facility serves as both a health clinic and a nursing school; nurses who complete the Kalungi program are deployed into village hospitals. The hospital has a 7 person staff with one medical assistant. The clinic sees 20-30 patients daily; this number increases to about 50 during malaria season. Some patients can pay a small fee of roughly 400 shillings, or 22 cents.

The project at Kalungi Hospital involved the electrification of the hospital, as well as the provision of clean water to the hospital and local community.

School children from the Kalungi community gather clean water.

A preliminary assessment was completed in the initial phases to determine the energy demands and the appropriate technology to meet those demands. Due to insufficient wind resources and the high cost of fuel, a photovoltaic system (PV) presented the best available option to meet the power needs of Kalungi.

To meet the electrification goal, a 1.6 kW solar array was installed at the hospital to serve the electricity needs onsite. This includes 80 energy-efficient lights, a refrigerator and diagnostic equipment. To provide clean water, a 2.6 kW solar array was installed several kilometers away at a well site. This array powers a pump which pumps the water up a hill to a holding tank at the hospital. A pipeline runs back down the hill with spigots in several locations to provide clean water to the community.

The separate pipeline helps to ensure that the water traveling to the tank remains under pressure and uncontaminated.

To cover the cost of ongoing maintenance and operations, the Kalungi Hospital sold excess water to the surrounding community. The revenue was used to train two long-standing hospital employees to maintain the electrification and water systems. Funds were also used to hire a security guard to keep the electrification system secure from thieves at all hours. In addition, Solar Energy Uganda Ltd., the group responsible for purchasing the system, offered a 5-year warranty for the water pump and a 25- year warranty for the PV system.

## Electrification Project Specifications

System size: 1.6 kW Solar array (0.2 kW DC system and 1.4 kW AC system)

Load: 80 lights, refrigerator, and diagnostic equipment

Total Electrification Project Cost: $38,000 (including contractor and equipment costs, excluding administrative/overhead costs).

## Electrification Benefits

- Extended clinic hours, including services to patients throughout the night
- Continued study for nursing students at night
- Capability to sterilize medical instruments
- Fuel savings of approximately $25,000/year
- Power for diagnostic equipment such as microscopes
- Refrigeration of vaccines, medicines, and diagnostic supplies
- Extended computer use and capability
- Increased surgery load; better overall patient care locally
- Less likelihood of patient transport to Masaka District facility for in-depth care

Solar panels provide electricity to Kalungi Hospital.

## Water Project Specifications

- Provision of clean water for health clinic and community
- Powered by 2.6 kW solar array

**Key components:** 50,000 L double wall, concrete water tank at the health clinic, 2.7 km of pipeline, submersible pump, UV water purification unit. Hydrological surveys ensure water site basin will survive all dry seasons.

**Note:** Purification of water is completed in several stages: A 12 M deep well was dug and developed. From the well, a water passage was constructed, filtering water through two chambers. From the source, the water is pumped by a PV direct submersible pump to the main tank a total distance 2.2 km. It is then distributed throughout the hospital, to a school, and to the community.

**Total Water Project Cost:** $83,000 (includes contractor and equipment costs, excludes administrative/overhead costs).

## Water Project Benefits

- Clean drinking water for community (100,000 people): In addition to the spigots installed at the Health Clinic, three spigots were installed in the local community in order to give the local residents access to clean, potable water
- Less likelihood that the local spring will be contaminated by animals or people collecting water since the area has been secured and the communal spigots are more convenient
- Less likelihood of water-borne illnesses at the hospital, including dysentery
- Improved cleanliness and hygiene at the health clinic
- Better overall community health– the doctor in charge expects to see significantly reduced incidences of dysentery and other ailments transmitted through unclean water

## ACKNOWLEDGMENTS

We would like to express our deep appreciation and gratitude to a number of individuals who shared their knowledge and experience in the production of this publication, especially Craig Cornelius, United States Department of Energy; David Kline and Peter Lilienthal, National Renewable Energy Laboratory; John Pitman, ATA Services contractor to Centers for Disease Control and Prevention; Solar Light for Africa; Sam Dargan, Great Lakes Energy; and Global Environment & Technology Foundation. We would also like to acknowledge the significant insight and contributions shared by other experts and stakeholders.

## FOR MORE INFORMATION

Listed below are additional resources for a more technical, in-depth understanding of electrification options.

**General**

Renewable Energy for Rural Health Clinics, National Renewable Energy Laboratory,

September 1998.

http://www.nrel.gov/

**Renewables for Sustainable Village Power, National Renewable Energy Laboratory.**

http://www.rsvp.nrel.gov/

**Photovoltaic Systems**

Installing Photovoltaic Systems: A Question and Answer Guide for Solar Electric Systems, Florida Solar Energy Center,

May 1999.

http://www.fsec.ucf.edu/

**Wind Turbines**

Small Wind Electric Systems: A U.S. Consumer's Guide, U.S. Department of Energy, March 2005.

http://www.eere.energy.gov/

**Hybrid Systems**

Economics and Performance of PV Hybrid Power Systems: Three Case Studies, Sandia National Laboratories, July 1998.

http://www.sandia.gov/pv/docs/PDF/ecnmcs3cs.pdf

**Batteries**

Lead-Acid Battery Guide for Stand-Alone Photovoltaic Systems, developed for the International Energy Agency, December 1999.

http://www.oja-services.nl/iea-pvps/

Batteries and Charge Control in Stand-Alone Photovoltaic Systems: Fundamentals and Application, Sandia National Laboratories, January 1997. http://www.fsec.ucf.edu/pvt/resources

Reciprocating Engines (Diesel, Gasoline, Natural Gas)

Technology Characterization: Reciprocating Engines, developed for the U.S. Environmental Protection Agency by the Energy Nexus Group, February 2002.

http://www.epa.gov/

For USAID Missions and other U.S. Government programs interested in retrofitting health facility infrastructure, the USAID Energy Team in the Bureau for Economic Growth, Agriculture and Trade, Office of Infrastructure and Engineering is available to provide additional assistance. Contact information is provided below.

***Gordon Weynand***

Energy Team Leader
Office of Infrastructure and Engineering
Bureau for Economic Growth, Agriculture and Trade
U.S. Agency for International Development
1300 Pennsylvania Avenue, NW
Washington, DC 20523
Phone: +1 (202) 712-4169
GoWeynand@usaid.gov

## End Notes

[1] To find wattage information for a given device, check the user's manual or it may be written on the back of the device.
[2] The typical daily "on-time" for a standard refrigerator compressor is 5-6 hr/day.
[3] Refer to the following for solar resource data: http://eosweb.larc.nasa.gov/sse/
[4] Refer to the following for wind resource data: http://www.wwindea.org/home/index.php
[5] For more information on the HOMER model see http://www.nrel.gov/homer. This model can be downloaded free of charge from that web site.

In: Power on
Editors: William R. Thomas, et al.
ISBN: 978-1-61209-202-7


*Chapter 2*

# POWERING EDUCATION OPTIONS FOR IMPROVING ENERGY SERVICES AT EDUCATIONAL FACILITIES IN SENEGAL*

***United States Agency for International Development, (USAID)***

## EXECUTIVE SUMMARY

In late June-early July 2009, the USAID Energy Team (EGAT/I&E/E) responded to a request by the USAID Senegal mission to assist its education program. USAID/Senegal is supporting the Government of Senegal (GOS) to construct new middle schools around the country, and some of the sites are in locations not currently connected to an electric grid. To date, 57 schools have been constructed under the PAEM program, and more schools are expected to be built in the future (under a different contractual framework). USAID/Senegal seeks a strategy on how to meet the energy needs of these schools, in order to ensure a quality education for students and inform future USAID programming in the education sector.

* This is an edited, reformatted and augmented version of United States Agency International Development, dated October 2009.

Energy Team staff, together with a local and international expert on renewable energy, conducted a two-week assessment trip to five different regions to collect data upon which the team could base recommendations. The team visited 4 schools from the USAID/PAEM program, and met with staff from the Ministry of Education; the PAEM program, other donors; and private-sector energy service companies in Senegal. The team was able to visit a school in Soum with a solar system installed through the PAEM program, and spent considerable time reviewing the plans for installation of another 8 solar systems. Detailed recommendations on these plans have been provided in a separate report to the USAID COTR and PAEM Chief of Party.

## Findings and Recommendations

Based upon the fieldwork and discussions noted above, the Energy Team has the following recommendations for USAID/Senegal to incorporate in future education programming:

- **Energy should be a criteria in the selection of equipment supplied to schools:** Although USAID's implementing partner and the GOS seem to have agreed upon a standard package of electrical equipment to be delivered under the PAEM program, little thought has been paid by either party to the electricity demand of this equipment, since most schools within the program are connected to the electric grid. However, Senegal's grid frequently suffers power cuts and voltage fluctuations. In the future, USAID must take the availability of electricity into account when designing equipment packages for schools. Given the cost of on-site electricity generation, off-grid schools should be provided only with equipment deemed critical, and the most energy-efficient models should be procured. Schools on-grid would also benefit from energy efficient equipment; expensive/important equipment should also be paired with components that protect against power surges/cut-offs and voltage fluctuations (i.e., UPS units for computers).
- **Experts should be enlisted to assist in the design of an energy strategy, as well as selection and installation of energy equipment:** Energy (especially alternative energy) is a highly specialized area, and expert advice should be solicited before strategies are crafted and acted upon, even if energy represents a relatively small piece of a

project. Before the Energy Team's assessment, the mission had made no provisions to contract expert advice centrally, nor has it insisted/budgeted for its education contractors to enlist such advice. As a result, education specialists who lack the requisite knowledge and training are making energy-related decisions with negative consequences.

- **Seek local assistance to the extent possible:** Expert assistance may be solicited in- country or abroad, but it is always desirable to secure expertise and equipment locally, if possible. Ideally, an expert will come up with the overall strategy and energy system design, and then help USAID, the GOS, or their partners procure the necessary equipment and inspect the installation. Based on the U SAID Energy Team's assessment, Senegal possesses firms capable of doing solar design, procurement, and installation. Canadian CIDA is funding a UNICEF initiative to provide small solar systems to off-grid primary schools in Senegal, and may be a good resource for USAID to consult with. Similarly, the EU is funding a solar pumping program; both programs conducted tenders to solicit proposals from professional firms who are managing the installation and maintenance of the solar systems. The USAID Energy Team is also available to assist with further strategy development (beyond the basic strategy laid out in this document).
- **Consider all energy options:** Given Senegal's rural electrification plans, USAID and the GOS need to develop a strategy that will consider all possible options—grid connection, diesel generator, solar pv, and battery/inverter—to supply electricity to any given site before selecting which is most appropriate and cost-effective. Each decision should be based upon specific, standardized criteria (i.e., cost, need, availability of equipment/fuel, capacity to maintain/service equipment, etc.) and should not reflect a knee-jerk predilection for one type of solution over another.
- **Any provision of energy equipment must be accompanied by strategies that address enduser training and service/maintenance:** Any plan developed by USAID and the GOS to supply power generation equipment to Senegalese schools must be accompanied by a strategy to provide routine, periodic training to endusers on how to operate and maintain the equipment they receive, as well as a strategy to provide/fund more specialized service and maintenance. This is particularly important for systems involving batteries. NO energy

system is maintenance free. Generally speaking donor-funded programs tend to have a poor record when it comes to sustainability of energy systems, so this is an issue that needs to be addressed in both program planning and implementation. USAID/GOS should examine the feasibility of maintenance contracts for equipment provided under education programs.

The attached report provides more details on the above recommendations, focusing on options for schools for which grid connection is not technically or economically feasible. The report discusses 3 scenarios based on energy demand—high, medium, and low. The schools with low energy demand would need small energy systems to power basic computer labs. The medium-demand scenario applies to schools with a computer lab and electrical devices in the administration area. The high energy demand scenario encompasses facilities with more equipment and more hours of use.

Cost projections for various types and sizes of power-generation equipment are provided for each scenario, to help USAID/Senegal and the GOS plan for the next phase of the school construction program. A summary of these costs is provided in the two tables below.

**Table 1. Upfront Capital Costs and Levelized Cost of Energy of Various Systems.**

| System | Small 4,000 Whr/Day | | Medium 14,000 Whr/Day | | Large 29,000 Whr/Day | |
|---|---|---|---|---|---|---|
| | Init Cost | COE $/kwhr | Init Cost | COE $/kwhr | Init Cost | COE $/kwhr |
| **GRID CONNECTED SYSTEMS** | | | | | | |
| Grid / Inverter / Battery | 6,200 | 1.00 | 23,600 | 0.92 | 44,800 | 0.78 |
| **OFF GRID SYSTEMS** | | | | | | |
| PV / Inverter / Battery | 23,350 | 1.17 | 77,300 | 1.07 | 161,000 | 1.07 |
| Gen / Inverter / Battery | 17,000 | 2.92 | 44,800 | 2.76 | 65,400 | 2.20 |
| Generator Alone* | 6,000 | 10.60 | 12,400 | 12.76 | 12,400 | 6.16 |

The first table illustrates the upfront costs of purchasing and installing a system (Init Cost), as well as the levelized Cost of Energy (i.e., the cost in US dollars per unit of electricity produced by the system over an estimated 20 year lifespan. The COE includes initial capital costs as well as operations costs and periodic replacement of key components). These figures were derived from

HOMER, a modeling software designed by the US National Renewable Energy Laboratory. Data collected from Senegal was fed into the program, to make the costs as realistic as possible. Nevertheless, the figures are illustrative and should be used for planning purposes only. The importance is in the magnitude of cost of any given system RELATIVE to any other.

This table shows that solar PV systems have the largest upfront costs (many times higher than a generator), but the lowest levelized energy cost of any of the off-grid options. Adding a battery/inverter to a generator will increase the upfront costs, but decrease the long-term operating costs of generator-based systems. The grid-connected option shows approximate costs for a battery-based back-up system that can supply electricity to schools when the grid is down.

The second table removes the initial investment costs, and focuses specifically on the cost of energy. The "full cost" columns incorporate the approximate capital AND long-term operation costs for each system over 20 years. The "no initial costs" columns represent the costs of operating each system (i.e., those costs that must be borne by the enduser), but not the upfront costs of purchase and installation.

From this table it becomes clear that the electric grid will almost always provide the most economic cost of energy (these figures do not include connection costs), and grid connection should always be the first option explored to power off-grid schools.

**Table 2. Levelized Cost of Energy with and without Initial Equipment Costs.**

| | Cost of Energy (COE) USD/kWhr | | | | | |
|---|---|---|---|---|---|---|
| | Full Costs | | | No Initial Costs | | |
| System | Small | Med | Large | Small | Med | Large |
| Reliable Grid | .30 | .30 | .30 | .30 | .30 | .30 |
| Generator Only | 10.60 | 12.76 | 6.16 | 10.34 | 12.61 | 6.09 |
| Grid / Inverter / Battery | 1.00 | .92 | .78 | .74 | .64 | .57 |
| Gen / Inverter / Battery | 2.92 | 2.76 | 2.20 | 2.21 | 2.10 | 1.83 |
| Solar PV / Inverter / Battery | 1.17 | 1.07 | 1.07 | .19 | .13 | .10 |

Solar PV systems, despite their high upfront costs, have a comparatively low cost-of-energy over the long term, provided the batteries are maintained well. The grid/inverter/battery option is supplied to illustrate the cost of supplying back-up power to schools connected to the electric grid.

# 1. INTRODUCTION

In late June-early July 2009, the USAID Energy Team (EGAT/I&E/E) responded to a request by the USAID Senegal mission to assist its education program. USAID/Senegal is supporting the Government of Senegal (GOS) to construct new middle schools around the country, and some of the sites are in locations not currently connected to an electric grid. To date, 57 schools have been constructed under the PAEM program, and more schools are expected to be built in the future (under a different contractual framework). USAID/Senegal seeks a strategy on how to meet the energy needs of these schools, in order to ensure a quality education for students and inform future USAID programming in the education sector.

Map of the team's route.

Energy Team staff, together with a local and international expert on renewable energy, conducted a two-week assessment trip in Senegal from June 22 to July, 3 2009. The team, together with local PAEM project staff, conducted site visits to several schools and met with education officials in Soum (Fatick), Tambacounda, Fongolimbi (Kedougou), and Dakar. The map below shows the team's route.

A classroom block at a grid-connected middle school in Kaolack.

The PAEM schools are built to a standard design: 3 buildings in a U shape (1 administrative block in between 2 classroom blocks), and a latrine block.

The program is providing the following electrical equipment to schools on the grid:

- 7-8 desktop computers
- 1 laptop computer
- 1 printer
- 1 photocopier
- 1 video-projector
- 1 UPS
- 1 telephone/fax
- 6-7 power strips

It was not clear from discussions with PAEM staff if the same equipment package was being provided to schools off-grid (this is discussed further in later sections), but this is a critical issue for selection of power generation systems for the off-grid schools.

The primary goals of the field visits and interviews were:

1. Ascertain the energy needs of middle schools participating in the PAEM program (as representative of the needs of future schools);
2. Understand what measures were currently being taken to meet the energy needs of these schools;
3. Assess the various technological options available to meet the schools' needs; and
4. Gain a better understanding of the Senegalese market for on-site power generation equipment, to determine if local expertise is available to assist USAID/Senegal.

The team's findings are discussed in the rest of this report as follows:

- Section 2 focuses on energy demand, explaining the concept of "load" in order to highlight the importance of energy efficiency and provide a framework with which to size and cost alternative energy systems.
- Section 3 examines three real load scenarios (small, medium, and large) based on equipment being used in the schools visited during the fieldwork.
- Section 4 examines the costs of various technological options for each scenario.
- Section 5 addresses steps that the donor/user must take in order to assure the sustainability of renewable energy systems.
- Section 6 summarizes the team's conclusions and recommendations.
- Annex A contains more information on technological options.

## 2. The Importance of Loads and Energy Efficiency

In order to identify and price options for providing energy to educational facilities that are not connected to a reliable national electricity grid, one must first know the daily quantity of energy required by a given facility. The amount of energy required may be referred to as a site's energy "load." (Sometimes the equipment requiring energy may also be referred to as the load). The cost of providing energy to an off-grid site varies directly with the quantity of energy to be supplied. For this reason, the selection of electrical/electronic equipment supplied to schools by the Senegalese government or donor programs needs to be closely coordinated with the electricity supply (grid or off-grid) in order to minimize investment costs and ensure that the equipment provided is usable. This concept is not well-understood in the current PAEM program or by USAID education staff.

Generally speaking, when grid power is available, most people/facilities do not think much about the quantity of energy they are purchasing. They expect the electricity to be available when they want it, and know that the electric utility company (i.e., Senelec) will charge for it based upon the quantity they've consumed. The meter on the enduser's home or building records the amount of electricity that is consumed and the utility sends a bill to the consumer for these units. If the user needs more energy, to power a new piece of electrical equipment, for example, he just consumes the electricity and then pays for it in his monthly bill.

Sites that cannot be connected to a reliable grid may have to generate their own electricity. In these cases, the user cannot assume that his supply of energy is unlimited—he must pay careful attention to ensure that he is able to generate enough energy to meet his needs. Since the equipment needed to generate electricity is expensive, it is very important to keep electric loads as low as possible, in order to minimize upfront investment and ongoing operation costs.

## 2.1. What Are Watthours?

A watt is a measure of electrical power (power reflects the rate at which energy is used). Electrical devices are rated in watts, telling the user how much power the device will consume in 1 hour. Watthours, the product of watts and time (watts x hours = watthours), are a measure of energy. A kilowatthour (kWhr) is 1,000 watthours. Watthours or kilowatthours form the basis of all of the designs that deal with providing reliable energy to an off-grid facility.

For example, a light bulb that consumes 20 watts of power in 1 hour will require a supply of 100 watthours of energy if left on for 5 hours. (20 watts x 5 hours = 100 watthours).

Determining the total daily energy requirement of a facility involves an exercise in which one first identifies all of the devices requiring electricity, then multiplies the watts of each device by the hours it will be used to determine the amount of watthours required to use that device. Then all of the watthours for all of the devices are added together to come up with the daily energy requirement of the facility.

## 2.2. Calculating the Daily Energy Requirement in Watthours (Whrs)

An example of the watthours required in one day by a very small facility could look something like the following:

| Load | Watts | Hours | Whrs/Day |
|---|---|---|---|
| Light | 20 | 4 | 80 |
| Computer | 125 | 4 | 500 |
| Fan | 75 | 4 | 300 |
| Total | | | 880 |

The three devices listed in the above table, if required for 4 hours in a day, would require a reliable supply of energy in the quantity of 880 watthours each day. In a real facility, the load requirements can vary from something this conservative to a system that would easily require 10,000 to 40,000 watthours of energy per day.

The quantity of required watthours/day becomes the basis for all of the system design and costing that follows. Therefore, when considering the type of energy system to provide to an off- grid school, it is imperative to go through this process. Although the energy advisor (or sometimes equipment vendor) will conduct the actual calculations, the education staff must be closely involved, as they need to understand the capacity and limitations of the systems they ultimately will acquire.

Providing energy by non-traditional methods can get quite expensive, as the cost of supplying energy is nearly linear. So, the cost to provide a system that will supply 20,000 watthours of energy per day will be twice as expensive as a system that will provide 10,000 watthours of energy per day. Therefore, the exercise of determining the daily load is normally an iterative one, requiring that the energy advisor not only thoroughly reviews the loads to be sure none are missed, but also takes a hard look to be sure all possible economies have been incorporated. The process likely will require considerable interaction between the energy advisor, program planner/funders and education staffs, and it is not unusual to re-calculate the daily energy requirement several times as compromises are made between cost and necessity. To date, the PAEM program has not undertaken a sufficient load analysis.

## 2.3. Importance of Equipment Selection

The following table provides illustrative ranges of the power consumption of devices commonly found in educational facilities.

| Load | Watts |
|---|---|
| Fluorescent Tube Light | 10W to 40W |
| Compact Fluorescent Light | 7W to 22W |
| Fan – Small Wall Mounted | 40W |
| Fan – Ceiling Mounted | 75W to 150W |
| Desktop Computer – Efficient | 125W to 150W |
| Inveneo Desktop Computer | Varies – possibly 60W |
| Laptop Computer – Standard | 40W to 80W |
| Laptop – One Laptop per Child | 6W to 15W |
| LCD Projector – Standard | 300W |
| Photocopier | 1200W and up |
| Printer | 200W to 400W |
| Cell Phone Charging | 5W |
| Laboratory Microscope | 40W |

The power consumption figures listed above are all representative of typical pieces of equipment. When designing a system, it is best to know the actual power requirements of the equipment that will be utilized in the facility, since there can be significant variation from the low end to the high end in many of these categories. It is almost always cost-effective to purchase the most energy efficient products available, especially for use in off-grid sites. Although energy efficient products may cost more initially, their lower energy consumption will enable the purchase of a smaller energy system, and may also save on fuel and replacement costs.

A facility load assessment might start off with the following table:

| Qty | Load | Watts | Hours | Whrs/Day |
|---|---|---|---|---|
| 10 | Light | 40 | 6 | 2,400 |
| 15 | Desk Top Computer | 125 | 6 | 11,250 |
| 4 | Fan | 75 | 6 | 1,800 |
| 1 | LCD Projector | 300 | 4 | 1,200 |
| 1 | Photocopier | 1200 | 2 | 2,400 |
| | **Total** | | | **19,050** |

Since a renewable energy solution that can provide this amount of watthours/day would be quite costly, the team should consider ways to reduce the number of watthours needed each day, and conduct the exercise again. A second iteration that incorporates (a) changes to more efficient equipment, and (b) reductions in the number of hours the high energy-consuming equipment is used could result in the following assessment:

| Qty | Load | Watts | Hours | Whrs/Day |
|---|---|---|---|---|
| 10 | Light | **20** | **4** | 800 |
| 15 | **Lap Top Computer** | 60 | 6 | 5,400 |
| 4 | Fan | 75 | 4 | 1,200 |
| 1 | LCD Projector | 300 | 2 | 600 |
| 1 | Photocopier | 1200 | **0.5** | 600 |
| | Total | | | **8,600** |

The result of the energy review in this hypothetical example took the daily energy requirement from 19,050 to 8,600–or a reduction of more than half. The reduction was accomplished by changing the type of computer and light bulbs being used, and limiting the usage times for certain pieces of equipment (see bolded type in the table above), but the end result does not represent a significant change in the quality of service to be delivered. However, the cost of the system needed to supply energy for the equipment in the second example would be less than half that required for the first example.

In summary, the important points to remember are:

- Energy is expensive – especially when working with off-grid solutions.
- Energy is measured in watthours.
- Energy efficient equipment, though sometimes more expensive to purchase than conventional equipment, is almost always a good investment for off-grid sites.
- An energy expert can do a lot upfront to reduce the number of watthours required by a facility, and therefore the cost of the energy system required by an off-grid site.
- The USAID/GOS education team should work with an energy expert to come up with a mutually agreeable determination of daily energy requirements for Senegalese schools, in order to ensure that the future education program plans equipment purchases accordingly and that

relevant staff understands exactly what the recommended energy generation equipment will be able to power.

# 3. Load Scenarios and Cost Comparisons

Based on our analysis of the PAEM project, we've specifically selected three load scenarios which can provide a basis for the mission's future deliberations on energy system/electrical equipment packages. These three system sizing assumptions are then used to project costs of the various on-site generation systems identified as most feasible, and discussed fully in Appendix A. The representative energy usage scenarios--low, medium, and high--are based on actual conditions witnessed in the field. Of course, there are numerous possible variations on these scenarios, depending upon what equipment package USAID and the GOS ultimately select. For purposes of the illustration of costs, we will use these three.

## 3.1. Low Energy Use Scenario/Small System

In this system, the goal is for the school to be able to use 8 laptop computers, 8 hours/ day. The underlying assumption is that the laptops are used during the day, and charged at night. This scenario could apply widely to some of the more remote, off-grid schools.

**Low Energy Use Scenario (Laptop Charging only).**

| Area | Qty | Load | Watts Each | Hours of Day | Usage Night | Total Watt Hours | Total Conn Watts |
|---|---|---|---|---|---|---|---|
| Computer Lab | 8 | Laptop Computers | 60 | | 8 | 3,840 | 480 |
| **Grand Totals** | | | | | | **3,840** | **480** |

Therefore, we will use a daily estimated load of 4,000 watthours, which will require a relatively small energy system.

## 3.2. Medium Energy Use Scenario/Medium System

The medium energy use scenario adds a small administrative area to the "small" scenario. The assumption is that the computer lab still operates during the day, while batteries are charged at night. The administration area also operates during the day.

**Medium Energy Use Scenario (Computer Lab and Small Admin Area).**

| Area | Qty | Load | Watts Each | Hours of Usage | | Total Watt Hours | Total Conn Watts |
|---|---|---|---|---|---|---|---|
| | | | | Day | Night | | |
| Computer Lab | 8 | Laptop Computers | 60 | 8 | | 3,840 | 480 |
| Admin Area | 4 | Desktop Computers | 200 | 8 | | 6,400 | 800 |
| | 1 | Printer | 200 | 2 | | 400 | 200 |
| | 1 | Copier | 1200 | 2 | | 2,400 | 1,200 |
| | 1 | LCD Projector | 300 | 2 | | 600 | 300 |
| **Grand Totals** | | | | | | **13,640** | **2,980** |

We will use a daily estimated load of 14,000 watthours for this scenario, which will require a medium-sized energy system.

## 3.3. High Energy Use Scenario/Large System

For the highest energy use scenario, we will assume that the computer lab has desktop computers, and that some lighting is added so that the administration area or study hall can be used at night. Some of the equipment operates longer hours.

We will use a daily estimated load of 29,000 watthours for this high-use scenario, which will require the largest system. (Note that the watthour requirement of this system could be reduced significantly by swapping desktop to laptop computers, or by using desktop computers that are far more efficient, such as the Inveneo computers that are now being purchased and delivered in some instances in Senegal.)

**High Energy Use Scenario (Computer Lab, Admin Area, Night Use and Longer Daytime Hours).**

| Area | Qty | Load | Watts Each | Hours of Usage | | Total Watt Hours | Total Conn Watts |
|---|---|---|---|---|---|---|---|
| | | | | Day | Night | | |
| Computer Lab | 10 | Desktop Computers | 200 | 8 | | 16,000 | 2,000 |
| Admin Area | 4 | Desktop Computers | 200 | 8 | | 6,400 | 800 |
| | 1 | Printer | 200 | 2 | | 400 | 200 |
| | 1 | Copier | 1200 | 4 | | 4,800 | 1,200 |
| | 1 | LCD Projector | 300 | 2 | | 600 | 300 |
| Lighting | 8 | Compact Fl. Bulbs | 22 | 4 | | 704 | 176 |
| **Grand Totals** | | | | | | **28,904** | **4,676** |

## 4. Technology Options and Costs for Each Scenario

Based upon our site visits and assessment of locally available technologies, the most likely options to provide energy to off-grid schools in Senegal are:

- Connection to the national grid;
- Grid Connection with an Inverter / Battery back-up system;
- Generator, alone or with an Inverter / Battery System; and
- Solar photovoltaic (PV) system with Inverter / Battery

If the national or local grid is located within a reasonable distance of the school and reliably provides power for at least 6-8 hours per day, paying for grid connection will generally be the most economic choice. The cost will be based upon the distance that must be wired (the utility normally will provide a quote in cost/km for the connection); and any equipment necessary to make the connection (i.e., transformer, meter, etc). These costs must be judged against those of the other options noted above. The on-going costs of a grid-connected system are the monthly utility bills. As the cost of grid connection will vary, we did not include this option in our costing table. However, in cases where grid power supply is available but not continuously (i.e., it may only be available at night or during certain other time periods), a battery/inverter may be used to ensure electric power is available when needed. In this scheme, the grid charges the battery, which can then be used to provide power when the grid supply stops. This option is included in our costing estimates.

In cases where the grid is not available or extension too costly, our cost projection data covers three technology options:

- PV / Inverter / Battery
- Generator / Inverter / Battery
- Generator Alone

## 4.1. Cost Assumptions

The figures in the following tables were derived from HOMER, a modeling software designed by the US National Renewable Energy Laboratory. The software runs calculations based on assumptions of costs, resources, and economics. Although actual current data from Senegal was fed into the model, any adjustment to the assumptions used (i.e., to interest rates, future fuel costs, etc) could have a significant impact on the long-term cost of energy for any given system. For this reason, the figures in these tables should be viewed as illustrative only, as costs of equipment and fuel change frequently. The significance is in the magnitude of cost of any given system RELATIVE to any other.

The assumptions/data used in the model include:

- **Equipment costs:** cover all major energy system components, wiring/cabling between the components, and installation. We have used cost data obtained from Senegalese vendors in Dakar to feed the model. For PV, the systems are based on conditions in southern Senegal, but there will be system variations in different locations throughout the country depending upon local climatic conditions and solar insolation. Equipment costs do NOT include allowances for USAID costs, subcontract costs or other management-related costs that are not within the control of the equipment suppliers.
- **Resource costs:** include fuel costs, i.e., the price of diesel and electricity from the grid. During our visit, diesel prices were

approximately \$1.50/liter at the pump. Costs to deliver fuel to the site vary widely, but in our experience it is not unusual for these costs to double the pump price. Therefore, we have used a diesel cost of \$3.00/liter. This includes the cost to purchase the fuel, load it into barrels, load the barrels into a pickup truck, transport and off-load the barrels at the site, re-load the empty barrels and return to town. For grid-supplied electricity, we used an average billing cost of \$0.30/kWhr (Senelec charges different prices to different consumers; this is neither the highest nor lowest).

- **Lifetime of equipment and system:** the cost comparison is run for 20 years, the lifespan of the PV panels. The life of the batteries is determined by HOMER and the manufacturer's data; typically 5-7 years. The system includes the replacement cost of the batteries when they are expected to wear out in the simulation. The inverter life is presumed to be 10 years. Generator life is estimated at 15,000 operating hours.
- **Interest rates:** we have based the cost of energy calculations on an investment interest rate of 2%.

## 4.2. System Costs

The following table shows the estimated upfront cost of each system size in each scenario, as well as the level ized Cost of Energy (COE). The COE is the cost in US dollars per unit of electricity that is produced by the system. It is a weighted average over the 20 year presumed life of the system. This figure can be compared against the assumed \$0.30/kWhr cost of electricity from the national grid.

The figures for the generator COE therefore are likely realistic estimates of the actual cost of energy that would be experienced by these sites.

Some general observations can be made from this table:

1. In general, if grid electricity is available at the site for a minimum of six hours, it usually makes sense to rely upon it. If fewer hours are available, the site may require additional energy generation or storage equipment (i.e., the battery-based back-up system shown above), which will bring costs up to the level of some off-grid solutions.

2. Solar PV systems have the largest upfront costs (many times higher than a generator), but the lowest levelized energy cost of any of the off-grid options.
3. Generators have the lowest upfront costs of the off-grid options, but their COE is much higher than for the other systems. A large savings in COE can be achieved by adding an inverter / battery system, which enables the user to operate the generator fewer hours (therefore, less fuel and maintenance is required).

**Upfront Capital Costs and Levelized Cost of Energy of Various Systems.**

| System | Small 4,000 Whr/Day | | Medium 14,000 Whr/Day | | Large 29,000 Whr/Day | |
|---|---|---|---|---|---|---|
| | Init Cost | COE $/kwhr | Init Cost | COE $/kwhr | Init Cost | COE $/kwhr |
| **GRID CONNECTED SYSTEMS** | | | | | | |
| Grid/Inverter/Battery | 6,200 | 1.00 | 23,600 | 0.92 | 44,800 | 0.78 |
| **OFF GRID SYSTEMS** | | | | | | |
| PV / Inverter / Battery | 23,350 | 1.17 | 77,300 | 1.07 | 161,000 | 1.07 |
| Gen /Inverter/ Battery | 17,000 | 2.92 | 44,800 | 2.76 | 65,400 | 2.20 |
| Generator Alone* | 6,000 | 10.60 | 12,400 | 12.76 | 12,400 | 6.16 |

Notes: (*): This does not include the cost of grid connection. (**) Only certain sizes of generators were modeled here. For the small system a 5 KW generator was used. For the medium and large systems, a 13KW generator was used. With perfect design, smaller generators could be used in all cases, but typically larger generators are provided to accommodate for increased future load or for other reasons.

## 4.3. Simple Payback in Years

The big difference between the systems is quite obviously the large initial cost of doing anything other than using a generator. All other options cost more on the first day of operation, but result in significant savings in the long-term cost of energy.

Of course, this means that if the site being considered is only going to be around for 1 year or less (or will be connected to the grid soon), then perhaps the generator IS the best option. To take a deeper look at the different options, we can look at the simple payback for some of the system examples. While this is not an exhaustive economic analysis, it does give some perspective to the choices that are left to the system designer.

Looking only at the Off-Grid solutions for the high energy use scenario of 29,000 watthours (or 29 kWhr) per day:

To switch from the Generator Only option to the Generator / Inverter / Battery system, we would have to spend an additional $53,000 in initial cost. However, we would save almost $4.00 / kwhr in energy costs. At 29 kWhr/day, annual consumption totals 10,585 kWhr/year. The cost savings in one year would be $41,917. ($3.96 / kWhr x 10,585 kWhr). The simple payback for switching to an inverter / battery / generator system in this example would be $53,000 in additional cost / $41,917 in energy savings per year, or 1.26 years.

To switch from the Generator / Inverter / Battery system to a system that provides most of the energy from PV, similar calculations would result in an approximate payback of 8 years.

To compare the PV system to using Generator only, the calculations would yield a payback of about 2.5 years.

Therefore, determining the optimal system for a given site should include some assessment regarding how long the site will be in operation. Other factors will influence payback calculations as well, particularly the estimates for future fuel costs, the cost of generators to be purchased in the future, and estimates for the life of batteries (see section 6 for more on batteries)

## 4.4. The Perspective of the Recipient

In order to view the cost of various systems from the perspective of the recipient (who will not bear the cost of system purchase in a donor-funded project), we have calculated the cost table again, eliminating the initial costs for equipment and installation. This leaves the ongoing costs of fuel, maintenance, equipment replacement, etc. Often the responsibility of paying for ongoing costs (particularly for fueling a diesel generator) is not adequately accounted for in donor-funded projects, resulting in quick failure of the system. (This is discussed further in Section 5).

In the solutions where the generator is the prime energy provider, there is little change in the long term COE regardless of who pays for the initial costs. This is because most of the costs are in the fuel and maintenance, which are the responsibility of the recipient.

**Upfront Capital Costs and Levelized Cost of Energy of Various Systems.**

| System | Small 4,000 Whr/Day | | Medium 14,000 Whr/Day | | Large 29,000 Whr/Day | |
|---|---|---|---|---|---|---|
| | Init Cost | COE $/kwhr | Init Cost | COE $/kwhr | Init Cost | COE $/kwhr |
| **GRID CONNECTED SYSTEMS** | | | | | | |
| Grid / Inverter / Battery | | 0.74 | | 0.64 | | 0.57 |
| **OFF GRID SYSTEMS** | | | | | | |
| PV / Inverter / Battery | | 0.19 | | 0.13 | | 0.10 |
| Gen / Inverter / Battery | | 2.21 | | 2.10 | | 1.83 |
| Generator Alone* | | 10.34 | | 12.61 | | 6.09 |

But, when a PV system is provided by a donor and upfront costs are excluded from the PV system option, there is a drastic change in the resulting COE. From the recipient's point of view, the donor is essentially pre-purchasing energy for many years to come. The resulting COE is a function of the maintenance costs and the periodic replacement of batteries and inverters.

If the batteries are managed well, these ongoing costs should be far less expensive than the costs of fuel, maintenance and generator replacement for a generator-only system. Nevertheless, even PV systems require routine maintenance, and poor battery management can dramatically increase the costs (and reduce the functionality) of a PV system.

Regardless of which option is selected, it is imperative that arrangements be made to allocate appropriate funds (from fuel savings, government budgets, donor funds or other sources) to pay for the required ongoing supervision, maintenance, monitoring and training involved with any of these systems. Otherwise, the initial investment may well be in vain.

## 5. Sustainability Issues

In order to assure that an alternative energy system is able to perform as designed, USAID must ensure that adequate attention and commitment is paid to long-term sustainability issues. These include:

- Proper Maintenance
- Proper Operation
- Proper Monitoring

## 5.1. Maintenance

Almost all of the energy system solutions noted previously in this report include batteries and, often, a generator. Both of these equipment items require ongoing maintenance.

The maintenance requirements for generators are fairly well known, since generators have been part of remote energy systems for so long. Nevertheless, maintenance on generators is usually severely lacking. In order for the generators to last, the manufacturer's recommendations on regular maintenance procedures must be adhered to. Quite often, this is not done because there was no financial commitment either by the donor, host government, or recipient to budget for these ongoing costs.

The same applies to batteries. In general, batteries are less understood than generators. Some batteries require less maintenance than others, but no system utilizing batteries is "maintenance free"--regardless of marketing assurances that claim otherwise. Battery systems are expensive, and the proper operation and maintenance procedures can determine whether the batteries last 6 months or 6 years. (This has been a hard lesson learned for PAEM).

We have found that it is best, at least in the first year (s) of an installation, to subcontract out the monthly maintenance requirements of the generator and batteries. This stipulation can be included in tender documents, and may involve additional cost. The maintenance work should be combined with ongoing staff training, so that after a reasonable amount of time, on-site staff can take over the maintenance functions.

## 5.2. Training and Monitoring Requirements

On-site power generation systems are designed to operate under certain assumed conditions. As explained in Section II, these assumptions include:

- Loads to be connected to the system; and
- Time that the loads are to be on in a given day.

Once systems are completed, a common reason for failure is that the system was not operated as designed. If the energy system is designed to power 6 computers for 4 hours each per day, but the computers are left on 8 hours/day, and other loads are plugged in as well, the system is clearly going to fail. Yet often facility staff are unaware of the constraints of their energy systems. They are not trained adequately (if provided training at all) on how the system operates, and how their own behavior will impact the system's performance.

Since many of the new schools being built in Senegal are in relatively remote areas, it will be important to train on-site staff how to regularly monitor key energy consumption and generation variables. At a minimum this will include filling out daily logs. In some cases the data can be submitted electronically to a website maintained by the equipment vendor, to enable remote monitoring. The proper monitoring system will vary depending upon project location, system complexity, and staff, but some form of monitoring is always required.

## Conclusion and Recommendations

There is a large, recognized need to expand the quality of education services throughout Senegal. In most cases, the increased education services will include a need for electrical energy. There are many school sites throughout Senegal that either are not connected to the national grid, or are supplied with a grid connection that will not support daytime energy needs. Therefore, creative, alternative methods need to be developed to be able to provide energy under these off-grid, or sub-standard grid conditions.

Providing alternative energy to these sites will require a commitment from all of the parties involved to learn about the possible solutions, and to dedicate the time, energy and resources into designing and installing the appropriate systems. Energy system investments can be costly; therefore, thorough planning is a must before funds are committed. USAID, together with the Government of Senegal and other partners, must first determine which energy needs are critical so that there is a commonly understood and accepted load profile to guide energy system design and costing.

Bulk procurement will usually result in a lower per unit cost and/or better service package for any technology. USAID should consider separating energy investments from larger education projects in order to ensure that adequate expertise is utilized and the best possible prices are obtained. At a minimum, contract language should necessitate that implementing partners procuring energy equipment utilize the services of a qualified, experienced system designer and solicit competitive bids for equipment, installation, and maintenance. USAID and the Government of Senegal must also assure that the appropriate training and ongoing maintenance of alternative energy systems are ensured. Otherwise, the systems will not perform as expected and the expensive equipment will sit idle.

# ANNEXES A. TECHNOLOGY OPTIONS

There are many technical options available to provide power to schools in Senegal. The most technologically and economically feasible are covered here. (We did not investigate the feasibility of wind or micro-hydro, since they are so site-specific and unlikely to be applicable to many locations).

## A.1. Grid Connection

Connecting a school to the grid involves, at maximum, the following installations:

Minimum requirements

- Low voltage connection Transformer – School: $ 8,000/km
- Control board and metering device: $ 75/connexion
- Internal wiring of the school: $ 200/connexion

Possible additional requirements

- Medium Voltage cabling from grid to site: $ 11,000/km
- Medium/Low Voltage Transformer: $ 8,000/unit

Generally speaking the national utility company (Senelec) is in charge of everything but the internal wiring of buildings To obtain an estimate of connection costs, the customer must request an assessment from the utility. After connection fees are paid, the only ongoing costs are for electricity consumption, which is billed monthly. Prepayment meters have been introduced for clients who prefer not having to pay monthly bills.

## A.2. Grid with Inverter / Battery

In cases where grid power supply is available but not continuously (i.e., it may only be available at night or during certain other time periods), a battery/inverter may be used to ensure electric power is available when needed. In this scheme, the grid charges the battery simultaneously as it powers the loads. The battery can then be used to provide power when the grid supply stops. The following graphic illustrates the components of such a system. This option was run in HOMER and the costs are explained in the text.

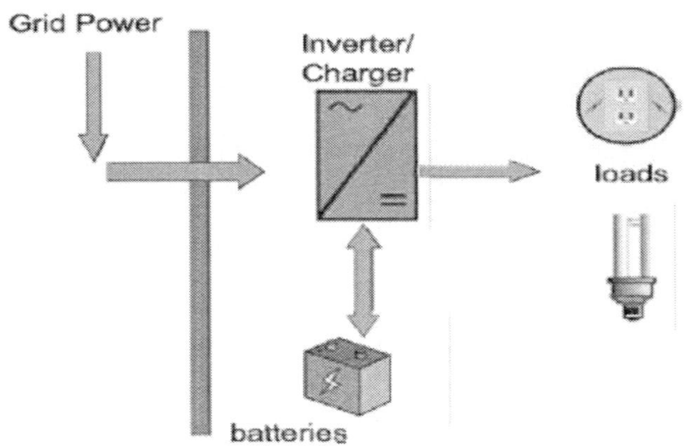


## A.3. Grid with PV and Inverter / Battery Option

This option applies to sites where the grid power supply is available, but with long discontinuities of the service. The idea is to generate power where the grid supply is cut for a long period. Therefore, the system is designed to limit the cost of storage and find the best compromise between "storage" and "alternative power generation".

The PV system and the grid system are both used to charge the batteries. When the grid supply is no longer present, the PV and the battery operate as a stand-alone power supply system. This system is unlikely to be economically feasible for USAID Senegal's education program given budgetary constraints, and so is not included in the report. However, we did run the calculations in HOMER and can provide the figures if needed.

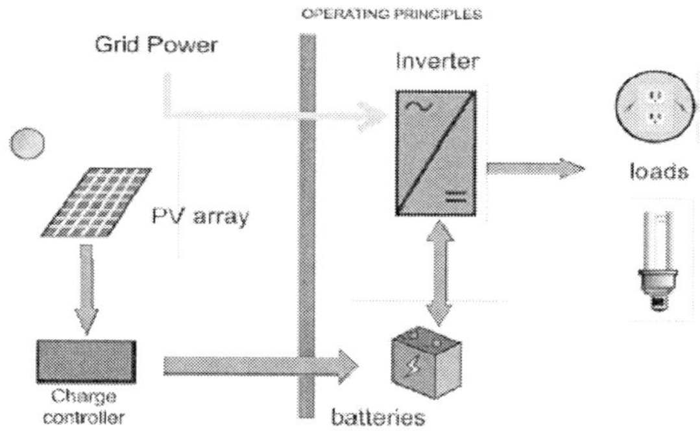


## A.4 Grid with Generator

This option applies to sites where grid power is available but suffers discontinuities of service. The generator operates as a back-up to the grid power supply, and stops as soon as the grid power supply recovers. We did not run this option in HOMER, as the battery/inverter option seemed a more plausible alternative.

## Generator-Related Options

## A.5. Generator

This option applies to sites where grid power supply is not feasible and where the loads do not require continuous supply. This option was run in HOMER and the costs are explained in the text.

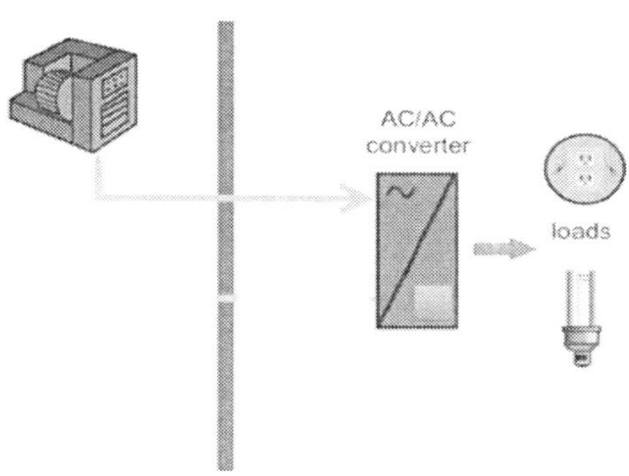

## A.6. Generator with Inverter / Battery

An inverter and batteries may be added to the generator, just as in the previous grid example. This option may be used at sites where grid connection is not feasible and the loads require long periods of supply. The generator charges the battery while simultaneously powering the loads. When the generator is off, the battery can be used to power the loads. In this situation the generator can be run at a greater efficiency, and for less time. The generator will therefore require less fuel and less maintenance over its lifetime than in the previous example. This option was run in HOMER and the costs are explained in the text.

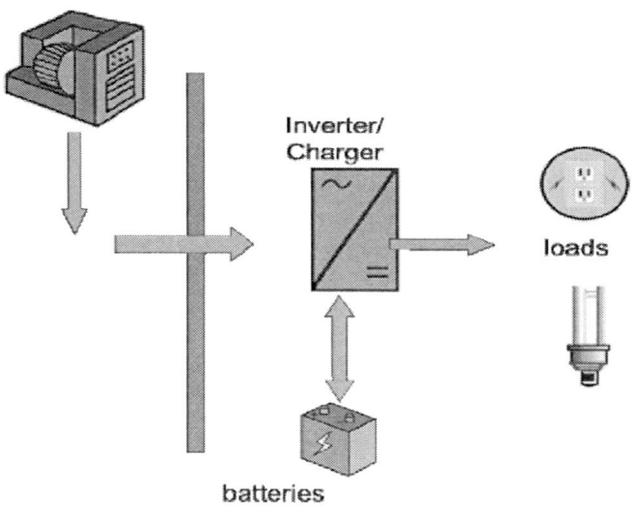


## A.7. Generator with PV and Inverter / Battery

This option adds a PV system to the system shown in option 6. The PV system ensures that the battery is always charged. The genset may be used to charge the battery whenever needed and may power the load simultaneously. When the generator is off, the PV system and/or battery can power the loads. Due to the cost of this system and the relatively simple power needs of Senegalese schools, we did not run this option in HOMER.

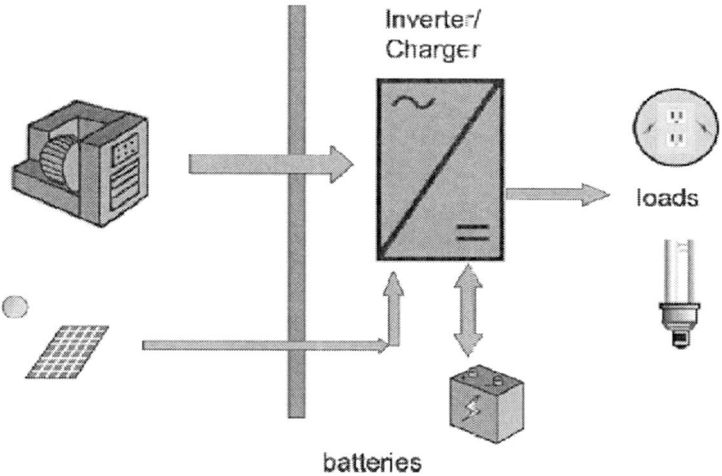


## A.8. PV with Inverter / Battery

This option suits sites where grid power is not feasible and which are isolated (difficult to supply fuel, expensive access for maintenance). This option was run in HOMER and the costs are explained in the text.

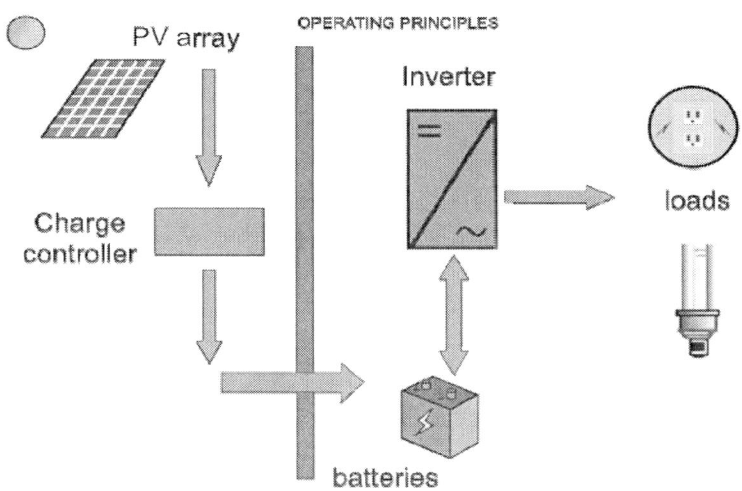

In: Power on
Editors: William R. Thomas, et al.
ISBN: 978-1-61209-202-7


*Chapter 3*

# POWERING TOURISM: ELECTRIFICATION AND EFFICIENCY OPTIONS FOR RURAL TOURISM FACILITIES*

***United States Agency for International Development, (USAID)***

"Tourism, in all of its forms, is arguably the largest industry in the world, having grown rapidly and almost continuously for the last 20 years. The growth of the international tourism industry has been particularly important to poorer countries. Between 1990 and 2000, international tourism receipts increased 133 percent in developing countries. During the current decade (2000–2010), the World Tourism Organization (WTO) forecasts that the number of international visitor arrivals will grow five times more rapidly in developing countries than in more established destination markets. "

*USAID and Sustainable Tourism: Meeting Development Objectives* June 2005

* This is an edited, reformatted and augmented version of United States Agency International Development, dated June 2005.

# POWERING TOURISM: AN INTRODUCTION

Rural tourism is a rapidly growing sector of the global tourism industry. Since rural tourism operations are often in remote locations, ensuring a reliable supply of energy can be a special challenge. Therefore, planning for reliable energy is vital and impacts all decisions made at rural tourism facilities—since these areas may have difficult access to fossil fuel supplies and no access to grid electric power.

Often a rural facility must be its own energy service provider, relying on energy resources available at the site and/or fuel transported to the site. Its staff must be knowledgeable in the day-to-day operation of power generation equipment, sound efficiency practices to reduce overall use of energy, and access to qualified technicians for periodic maintenance and equipment repair. Unlike tourism facilities with access to grid power and other utility services, rural off-grid facilities must bear the full cost of operating reliable on-site power and water systems—which typically result in power costs that are more expensive than electricity provided by a grid. Since rural facilities must be good stewards of the land, they face the added pressure of producing energy by the most sustainable means possible, and using energy as efficiently as possible.

Powering Tourism provides information to help tourism professionals understand and evaluate the range of options for meeting their off-grid energy needs, highlighting efficient and sustainable use of energy. The guide provides a seven-step approach to guide the reader through the process of creating a reliable and affordable energy system by working with staff and visitors, equipment vendors, installers, financial institutions, and energy service providers. Powering Tourism illustrates each step through hypothetical and real life case studies. Transportation needs are not covered in this guide, which focuses on applications requiring electricity and heat.

## Energy and Rural Tourism

The term "rural tourism" covers a number of different facility types, locations, and styles. Facilities may include museums, welcome centers, and park and lodging accommodations ranging in style from the most basic to luxurious. Many are designed as sustainable tourism facilities, and all are in remote areas. Sustainable tourism is defined as a set of principles and practices that involves environmentally- friendly operations such as minimizing consumption of non-renewable resources, bringing tangible economic and social benefits to local communities, and participating directly in cultural and natural heritage conservation efforts. [1]

Rural tourism facilities are found in all corners of the globe—from deserts to tropical and temperate forests and from coastal beaches to high mountains. A resort in an equatorial climate with little seasonal weather variation will have more consistent energy needs throughout the year than one in a polar region with extreme seasonal temperature variations. The location not only determines how accessible electricity and fuel resources are, but also what renewable resources are available for solar-, wind-, or water-powered generation equipment. Additionally, how and where the facility is sited affects energy requirements. Those facilities that take advantage of environmental factors such as shading in hot climates, use of natural light, air flow, and other factors can reduce the need for additional energy.

Small facilities typically have lower energy requirements than large ones, but ultimately, the amount of electricity and heat consumed will vary directly with the type of services provided. Within the context of rural tourism, some facilities provide only the most basic services with minimal energy requirements (i.e., visitor centers), while others are more luxurious, offering air conditioning, hot water, and electronic entertainment.

As shown above, virtually anything—type, style, location, climate, staff, guest levels, behavior, and operating hours—can affect energy demand and consumption at tourism facilities. Energy may be required for lighting; to prepare and store food; to pump, purify, and heat water; to provide entertainment; to charge batteries; and to maintain comfortable indoor temperature and humidity levels for staff and visitors. Further, as the facility evolves, so will its energy needs. The following sections will help guide you through the process of taking all of these factors into consideration to obtain the right energy system for your facility.

## A Stepwise Approach to Designing an Energy System

A tourism facility's energy system consists of:

- Devices that require electricity and/or heat.

- On-site power and heat generation equipment.
- The people who use the appliances and operate the power equipment.

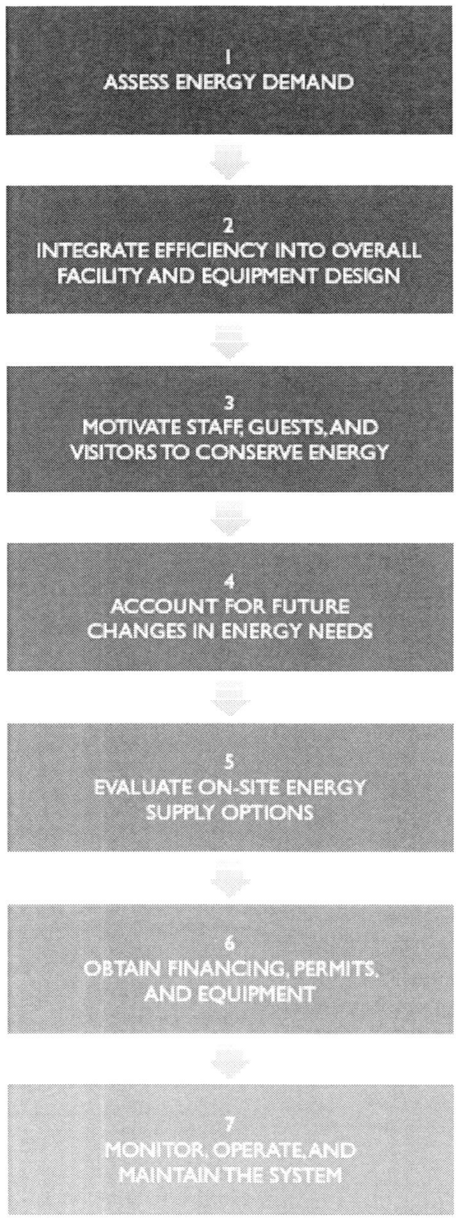

A well-designed energy system minimizes energy waste. Applying an organized approach to energy system design will help ensure that the system sustainably and reliably meets energy needs over time—even as a facility grows, people's behavioral patterns change, and equipment ages. The seven-step approach described in the following sections covers all aspects of planning, installing, and managing an energy system, from identifying energy needs to procuring efficient appliances and selecting appropriate on-site power generation equipment. This guide is technology-neutral, but highlights factors such as cost, reliability, environmental impact, and maintenance and operations issues for each power generation option. This allows readers to choose the system which best satisfies their needs.

The stepwise approach is intended as a guide for planners, managers, and other decision makers who will be relying on the advice and expertise of qualified energy practitioners, finance professionals, and additional service providers to perform the tasks in each step. This guide is not designed to serve as a how-to manual—our goal is to provide you with the information you need to make informed choices regarding energy supply decisions for your project. Case studies throughout the document provide examples of how actual facilities generate and use energy—note that these "real life" examples of energy use may not necessarily represent "best practices" or the most sustainable options available. The appendices provide tools that can help you put some of the suggestions described in the steps into action.

## The Paradise Resort

Throughout the guide, the Paradise Resort serves as a hypothetical case study that illustrates energy planning, efficiency, use, and design for a facility in Steps 1 – 5. The energy need and use estimates are based on a sample remote tourism facility (see full description below); costs are in US$ and estimates are modeled using the HOMER software developed by the National Renewable Energy Laboratory (www.nrel.gov).

The Paradise Resort is located in a tropical climate and has six bungalows for guests, as well as a main lodge containing a kitchen, dining facilities, and reception and office areas. There is lighting inside the lodge and bungalows, but minimal outdoor lighting at night. Heating requirements are only for cooking, for which the restaurant uses a combination of propane and wood. Paradise Resort has a small circulation pump that pumps water from a nearby source (gravity-fed) to a tank on the roof of the building.

## Step 1. Assess Energy Demand

The first step in the selection of an energy system is to assess energy demand. For existing facilities this will require an energy audit. For new facilities in the planning phase, the energy assessment will estimate future energy use. The assessment should include an estimate of the amount of energy required, a projection of how much it will cost, and identification of opportunities to improve energy efficiency and reduce energy consumption.

### Factors that Affect Energy Demand in Tourism Facilities

- The size of the facility, measured in number of rooms or visitors.
- Total number of employees and number of employees lodged at the facility.
- Climate.
- Annual operating schedule of facility.
- Level of service.
- The type of equipment and appliances operating at the facility.

Tourism facilities typically use energy for lighting, communications, indoor space heating and cooling, water pumping and treatment, and the operation of office equipment. Facilities that offer food and lodging services may also require energy to prepare food, heat water for bathing and cleaning, provide laundry services, and power equipment for guest entertainment. These activities require energy in the form of either electricity or heat.

Specific information needed to assess energy demand includes: (1) an inventory of electrical devices and how often they are used; and (2) an estimate of heating/cooling and water requirements.

### Quantifying Energy Use

A thorough energy audit should be conducted by a skilled professional. Nonetheless, this guide provides a simplified tool (Appendix 1) that readers can use to get a rough understanding and estimate of their energy demand. The tool relies on two primary pieces of information—equipment power ratings and energy usage.

The power rating, measured in watts, indicates the rate at which a device converts electrical energy into another form of energy (such as light, heat, or motion). Different units are used to quantify power and energy.

The amount of energy a device uses depends on the power level at which it operates, and the length of time it operates. Watts are a power measurement—either power capacity, or power requirement. Devices that use electricity are measured in watts (W); power production is usually measured in kilowatts (kW) (a kilowatt is equal to 1,000 watts) or megawatts (MW) (a megawatt is equal to 1 million watts). Energy use is obtained by multiplying a device's power rating (in watts) by the number of hours it operates to arrive at watt-hours (Wh). For example, a 50 watt compact disk CD player operating for one hour will consume 50 watt-hours. Watt-hours are usually summarized in Wh/day, meaning how many hours a day the device is estimated to operate and how much energy it will use during that time. Other measurement terms include Btu (British Thermal Units) and joules, which measure heating and cooling requirements and also the energy content of fuels. For the purposes of this guide, power requirement measurements will be in watts to show power needed for a device; energy measurements will be in watt-hours (Wh) to show energy needed to use that device for a certain amount of time.

## Electrical Appliance Inventory at the Facility

To determine the current electricity requirement of the facility, make a list of all electrical appliances, and note the power specifications from the appliance's label or user manual (usually in watts). For future requirements, an estimate of power specifications can be based on a growth plan for the facility (see Step 4). Appendix 2 provides power ratings and average energy usage of various energy- consuming devices, on an annual basis, for both efficient and inefficient models. It is important to include the number of outlets available for guest/staff use (for camera charging, electric razors, etc).

The size (in terms of the wattage that can be accommodated) and number of outlets is directly related to the type/level of service a facility wishes to provide, and can have a significant impact on energy use.

## Heat Requirements

In addition to energy used for electrical devices, energy in the form of heat is a requirement for cooking, water heating for bathing and cleaning, and space heating. For cooking, a facility's energy requirement is determined by the number and type of meals prepared. Kitchen staff can provide estimates of cooking fuel requirements based on meals served. Hot water needs for bathing and cleaning can be calculated using estimates of daily hot water volumes and temperature. Space heating requirements are calculated using room volumes, room insulation characteristics, outdoor temperatures, and desired indoor temperatures. Because electrical heating has such a high energy requirement, heating is almost never an efficient application for renewable energy in electrical form— ideally, heat should be supplied from the burning of various fuels (i.e., natural gas) or through solar heaters.

## Water Requirements

For facilities that provide food and lodging services and do not have access to city water services, water pumps can be among the largest energy consumers.

Three factors determine the quantity of energy required for water pumping:

- The height that the water must be raised by the pump and the amount of energy it takes to transport the water to the site if it does not use gravity.
- Daily quantity of water required by the facility.
- Water purification needs—water pumps must be powerful enough to move water through filtration systems. Certain water purification systems, such as ultraviolet or desalination, might also have additional electricity requirements.

An energy assessment report or audit will include a set of values that represent estimates of the facility's energy needs. As shown in the following example, estimates often will be given in Wh/day, which is a useful measure for comparing the energy consumption of different facilities, or for assessing progress toward meeting energy efficiency goals for a single facility. This calculation will help identify areas for improving energy efficiency, whether by investing in more efficient equipment, or changing energy use behavior of staff and guests. It will also help account for potential changes in future energy consumption.

| Area | QTY | Load | Watts Each | Total Watts | Hours per day | | Total Watt Hours |
|---|---|---|---|---|---|---|---|
| | | | | | Daytime | Night | |
| Lodge Rooms | 6 | Lights (general) | 60 | 360 | | 8 | 2,880 |
| | 6 | Lights (bathrooms) | 60 | 360 | | 2 | 720 |
| | 6 | General Purpose Outlet | 75 | 450 | | 4 | 1,800 |
| | 6 | Fan | 150 | 900 | | 8 | 7,200 |
| Kitchen | 1 | Refrigerator | 100 | 100 | 12 | 12 | 2,400 |
| | 1 | Freezer | 100 | 100 | 12 | 12 | 2,400 |
| | 4 | Lights | 60 | 240 | | 8 | 1,920 |
| | 4 | Misc. Appliances | 100 | 400 | 4 | 8 | 4,800 |
| General Lodge Area | 6 | Lights (guest area) | 100 | 600 | | 8 | 4,800 |
| | 6 | Lights (office area) | 100 | 600 | | 4 | 2,400 |
| | 2 | Computers for Guests | 150 | 300 | 4 | 4 | 2,400 |
| | 2 | Computers for Lodge | 150 | 300 | 8 | 4 | 3,600 |
| | 1 | TV / Entertainment | 200 | 200 | 4 | 4 | 1,600 |
| | 1 | Satellite System | 80 | 80 | 12 | 12 | 1,920 |
| | 2 | Fans | 150 | 300 | 4 | 4 | 2,400 |
| Pumping | 1 | Small Circulation | 500 | 500 | 4 | 4 | 4,000 |
| Site Area Lighting | 6 | Outside Lights | 40 | 240 | | 12 | 2,880 |
| Grand Totals | | | | 6,030 | | | 50,120 |

## Example 1. Paradise Resort Initial Energy Assessment

An initial assessment is conducted at the resort to determine the current electrical loads. As the chart below shows, the facility uses 50,120 watt-hours a day (Wh/day). To calculate Wh/day, multiply quantity of appliances by watts required for each by hours per day in use. The following example calculates Wh/day for the fans: 6 (number of fans) x 150 (Watts each) x 8 (hours in use per day) = 7200 Wh/day total. Wh/day is a useful measurement for total load requirement (power requirement and hours of use). In Step 5 when the system is sized, these estimates will be used to determine the total energy requirement, measured in kW (1,000 watts).

This example includes devices with fairly modest loads most likely found in existing facilities not utilizing energy efficient equipment. Energy requirements could change dramatically if additional equipment is added or the amount of time existing equipment is used changes. As shown in the graphic below, lighting uses a significant portion of the electricity consumed by Paradise Resort, followed by the fans and computers. The next step will look at energy efficiency options to decrease the 50,120 Wh/day requirement.

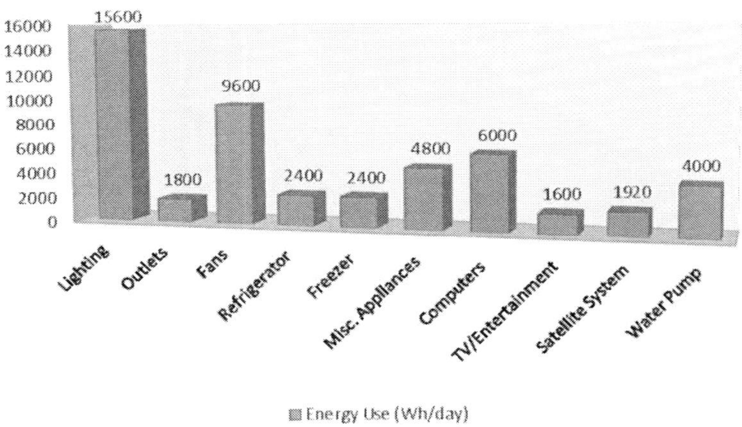


## STEP 2: INTEGRATE EFFICIENCY INTO OVERALL FACILITY AND EQUIPMENT DESIGN

Now that you have a preliminary estimate of your facility's energy demand, you can begin to consider methods to reduce that figure, and gain the resultant economic and environmental benefits. One of the key methods to improve the sustainability of any tourism project is to incorporate principles of energy efficiency. Energy efficiency should be a primary design objective for all new developments and, in particular, for tourism facilities that are built in remote and off-grid locations. Improving energy efficiency allows a facility to reduce energy consumption without altering the level of service or amenities provided.

Why is energy efficiency important for tourism operations?

- Integrating energy efficiency in a project can significantly reduce the initial cost and maintenance requirements of the facility's structures, appliances, and other end-use equipment. Buildings that are designed to be cooled year-round by natural cross-ventilation and ceiling fans can be built of light and often locally available materials, and do not require an airtight structure, wall insulation, and double-glazed windows and glass doors. Reducing the need for air conditioning units, which are expensive to purchase and operate, also reduces demand for qualified technicians.
- Purchasing energy-efficient equipment and appliances can be highly cost-effective, especially when electricity costs are high (as they are in all locations that must generate their own power). For example, replacing a 60W incandescent bulb with a 15W compact fluorescent lamp (CFL) in a grid-connected urban hotel generally can save enough over the life of the bulb to recover four to five times the initial investment. Likewise, replacing a showerhead that consumes 15 liters/minute with a low-flow model (consuming 8 liters/minute) saves considerable fuel and water and yields an annual return on investment of around 200 to 400 percent. Even though these returns are impressive, they represent only a fraction of the actual returns that can be achieved when making the same efficiency investments in remote facilities that self-generate power, since these sites often pay more than twice the market rate for the fuel used for their electricity generation.
- In an off-grid site, the capacity of the required energy generation systems is largely determined by the facility's projected energy consumption. Therefore, a facility that is designed with energy efficiency in mind is able to reduce the size, complexity, and cost of its energy supply and back-up systems.

Creating a facility that uses energy wisely and sparingly requires more than simply choosing and installing the highest efficiency appliances money can buy. It necessitates a wider approach that incorporates efficiency concerns in all aspects of the project, such as the siting and orientation of the buildings; the materials and architecture of the structures; the design and color schemes of interior spaces; and the equipment, appliances, and systems that provide comfort and meet the needs of visitors, guests and workers. Facility design/enhancement entails consideration of all components of the facility (including public areas, offices, kitchens, and/or staff quarters), not just those pertaining to visitors.

The information provided in Step 2 will help you become better aware of the various concepts, features, and choices that have a significant impact on achieving energy efficiency and sustainability.

## Buildings and Structures

The facility's architects or designers should be intimately familiar with the project location, as well as its climate and weather patterns, and use this knowledge to determine the optimal siting, orientation, and design of the structures. Site-specific features—such as predominant wind directions, topography, vegetation, and available water sources—can have a huge impact on how much energy is consumed by buildings and the level of comfort offered to occupants. Dependable breezes can cool buildings by cross-ventilation; trees can act as wind barriers or serve to shade roofs and walls; and a stream or spring located at a higher elevation can supply pressurized water by gravity alone.

Facility architects and designers should also be familiar with sustainable building concepts and be able to design efficient structures that can maintain a comfortable indoor environment with minimum energy input. Some of the typical features to be considered in designing energy efficient buildings are presented below.

- *Tropical climates:* Passive cooling strategies for buildings include: using well-insulated or highly reflective roofs to reduce the amount of heat that is radiated indoors; relying on overhangs or other features to shade walls and windows that are exposed to the sun; selecting light construction materials that have a low thermal mass for walls, roofs, and floors; and judiciously positioning windows and other openings to promote air flow through the buildings.
- *Temperate and colder climates:* Passive heating features include: maintaining a proper balance between floor area and the total window area on exterior walls that face the winter sun; using high thermal mass materials in walls and floors to absorb the heat of the sun during the day and gradually release it into the living space during the night; insulating roofs, attics, exterior walls, and foundations to improve

comfort (i.e., keep heated during winter and cool in summer); installing insulated windows and relying on shutters and tight-fitting insulated curtains to further reduce heat losses at night; and sealing the building envelope to minimize air infiltration.

Incorporating passive cooling and heating features in buildings does not necessarily result in more expensive structures. Palm thatch roofs have a low thermal mass, are naturally insulating, and can be built in most tropical regions with local material and labor. Straw bales are available at low cost in many parts of the world and can be used to build sturdy exterior walls with thermal performance that is two to three times better than that of well-insulated walls in modern buildings. Shutters and insulating curtains may be able to be made on-site at low cost, and can reduce heat losses through a double-glazed window during cold winter nights.

Finally, architects and designers should create structures that rely fully on natural sunlight for daytime illumination and resort to artificial lighting only at night. Even under overcast skies the average daytime illumination level in most regions is typically 15 times greater than that required to perform average indoor tasks.

## Lighting

The most commonly available energy efficient options for lighting applications in hotels consist of compact florescent lamps and linear fluorescent lamps, which on average consume four to seven times less energy than incandescent bulbs per unit of light output (lumen). Given that these efficient lamps can be used for virtually all lighting applications, they should be the obvious choice for interior lighting in facilities that run on expensive self-generated electricity.

The design of the lighting fixtures and the environment in which they operate are equally important in determining how much energy is required to produce the desired level of illumination in a given space. Ideally, fixtures should be able to transmit most of the light that is produced by the lamps, while the ceilings, walls, curtains, and other large furnishings should be light-colored so that they can reflect incident light back into the room rather than absorb it.

Fuel-based lighting—such as paraffin candles and oil, kerosene, and propane lamps—is often used in off-grid locations, mainly because it is cheap and does not require a source of electricity. However, these benefits come at a high cost, as fuel-based lighting is extremely inefficient and consumes a large amount of energy for the light it produces. In addition, there can be serious indoor air quality and resultant health issues for people with long-term exposure to fuel-based lighting.

## Water Conservation

Facilities that have easy access to abundant water sources often believe that water is free and treat it as such. In reality, however, unless it can be used directly as it falls from the sky, water is seldom free. Obtaining water from a source and transporting it to the point of use generally requires energy, infrastructure, and equipment (e.g., wells, spring catchments, pumps, pipes). Using this water requires more energy, infrastructure, and equipment, such as storage tanks, water treatment systems, pipes, water heaters, wastewater collection, and disposal systems.

### Using Refrigerators and Freezers

Energy to power refrigerators and freezers can account for a large portion of a tourism facility's energy needs. New, efficient models of electric refrigerators and freezers that use between 1 and 1 .5 kilowatt-hours per day are suitable for use with renewable energy systems. Self-powered refrigerators that burn propane or natural gas may also be a good option for some facilities. Here are a few tips concerning their use:

- Smaller capacity units use less energy than larger capacity units. Use the smallest size possible to meet facility needs.
- Keep door seals clean to minimize energy loss.
- Clean condenser coils at least once per year.
- Keep the unit in a cool place—out of direct sunlight and away from stoves and other heat producing appliances.
- Keep the unit full. Mass in the refrigerator will keep the cold in.

The energy cost of water is surprisingly high, especially for the hot water used in bathrooms and kitchens. Pumping and heating water can account for over half the total energy requirements of an efficient off-grid facility.

Water conservation opportunities can have a significant impact on reducing water usage. While conventional beach facilities consume 500 to 800 liters of water per guest- night (excluding irrigation water), efficient facilities need as little as 200 to 300 liters per guest-night, and some off-grid hotels operate with less than 100 liters per guest-night.

The devices and systems that hotels can use to minimize their water use include:

- Faucet aerators which, depending on the model, consume from 2 to 1 0 liters per minute. ▫ Low-flow showerheads which consume from 7 to 10 liters per minute.
- Self-closing faucet and shower valves.
- Pedal valves for kitchen and bar faucets to ensure they use water only when needed. ▫ Waterless or air-flush urinals.
- Water-saving toilets, micro-flush toilets, or dry composting toilets which operate well and use no water at all.
- Grey-water collection and reuse systems.
- Efficient washing machines that recycle part of the wash water they use and consume less than 1 0 liters of water per kilogram of linens processed.

Water conservation must also be included in facility operations. Employees should be motivated and trained to use water efficiently; guests should be invited and encouraged to participate in the facility's conservation efforts; water consumption should be tracked regularly to ensure this resource is in fact efficiently used; and water supply, storage, treatment, and end-use systems should be properly operated and maintained to minimize leaks and unnecessary losses.

Water—as well as energy, chemicals, and labor—can also be conserved indirectly by reducing the amount of towels, bed sheets, table cloths, and other linens the hotel sends daily to the laundry. Measures that can be put in place to achieve this important goal are simple, well-known, and highly effective when implemented correctly. They include adopting voluntary towel and bed linen reuse programs; reducing the frequency with which guestroom towels and bed linens are changed (typically every 2, 3, or 7 days, depending on the type of accommodation); and replacing tablecloths with runners, laminated place mats or, nothing at all.

## Energy-Efficient Equipment and Appliances

Efficient appliances perform the same work as less efficient appliances but require reduced energy to do so. The lower power rating of an efficient device is not an indication that it does less. Even seemingly small reductions in equipment power ratings can result in significant reductions in energy consumption.

Designers and engineers should be extremely careful to avoid over-sizing the appliances, equipment, and systems they select. Each year, tourism facilities waste billions of kWh of electricity operating devices that are bigger and more powerful than what is really needed. This includes guestroom electric kettles that can boil two liters of water when only one-third of a liter fits in a cup; conveyor toasters that continuously draw 3 to 4 kW of power even though they generally spend most of their time running empty (pop-up toasters work just as fast but use electricity only when actually toasting bread); and oversized water pumps that operate very inefficiently at a fraction of their design capacity. Thus, designers and engineers must first carefully gauge end-use needs and the capacity of the devices that are required to adequately meet them, and only then determine which devices can deliver the required capacity with the least amount of energy.

### Online Resources for Energy Efficiency Products and Practices

- ENERGY STAR and Energy Guide
  www.energystar.gov
- Federal Energy Management Program
  www.eere.energy.gov/femp
- Energy Saving Trust
  www.energysavingtrust.org.uk
- Minimum Energy Performance Standards:
  www.energyrating.gov.au
- EU Energy Label
  www.energy.eu
- Topten
  www.topten.info

High-efficiency alternatives exist for virtually all equipment, appliances, and devices, although identifying them through the barrage of manufacturer claims can be a challenge. This identification process can be greatly facilitated by relying on the information and recommendations provided by recognized energy-efficiency labeling organizations, such as the United States ENERGY STAR program, EnergyGuide, and Federal Energy Management Program (FEMP); the United Kingdom's Energy Saving Trust; Australia's Minimum Energy Performance Standards; the European Union's Energy Label; and Europe's Topten. Although there are differences among these labels, their objective is to rate products according to their energy use or provide minimum efficiency guidelines. This enables consumers to identify high-performance products and make informed decisions when choosing among various options. Many of these organizations also offer useful websites and detailed information on various efficiency issues related to the selection and use of appliances and other energy-consuming devices.

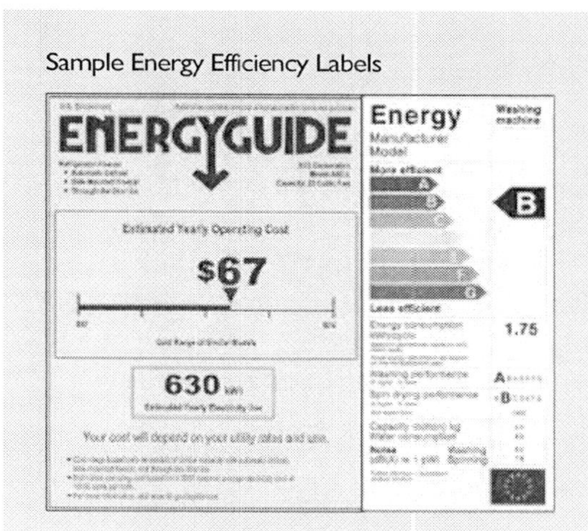


Energy efficient products, as a rule, are more expensive than their less efficient counterparts. However, the additional initial expense is typically recovered many times over by the energy savings achieved during the service life of the products. In the case of off-grid hotels, incorporating the use of a range of energy efficient products will result in direct savings up-front, because of the reduced costs of installing lower-capacity electricity generation systems.

Using high-efficiency products can achieve astounding returns on investment, especially in facilities that pay twice or more the market rate for self-generated electricity. Those responsible for selecting equipment and appliances must look beyond price tags and evaluate the life-cycle cost of the devices, and their impact on the facility's energy systems, or run the risk of paying dearly for hasty and ill-informed decisions. Life-cycle costs are a more accurate representation of the real cost of a purchase as they account for the initial capital costs, maintenance and operation costs, fuel costs, and the salvage costs over the entire life of the system. Ideally, cost estimates and vendor price quotes include a life-cycle cost estimate. For example, if a facility were investing in a new air conditioning unit, the cost estimate would include the cost of the unit, how much it would cost to run it each year, and how many years (estimated) it would operate. A planner could compare a conventional unit to a more efficient one to evaluate price and energy savings in the long term.

## Equipment Control and Automation

Operating lights and equipment when they are not needed is a common problem that can waste a lot of energy in facilities. The most cost-effective way to eliminate this problem is to make staff, visitors, and guests aware of the need to conserve energy, as well as train workers to incorporate energy conservation in their daily work routine. In areas where this low-tech approach is not effective, however, facilities can rely on a range of devices that are designed to automatically prevent lights and equipment from running longer than necessary. These include:

- Timer switches, which turn lights and equipment off after letting them run for a certain amount of time.
- Programmable timers, which automatically turn lights/equipment on and off according to a pre-determined schedule.
- Occupancy sensors, which are generally used to turn lights and fans on only when the space is occupied.
- Photocells, which allow lights to come on only when ambient lighting levels drop below a minimum acceptable threshold.

- Energy-saver switches, which turn off guestroom lights and fans soon after the guest pulls the card-key or key-ring from the switch as he/she leaves the room.

Although they can be highly effective in minimizing energy waste, these automatic controllers must be periodically checked to ensure they are correctly calibrated and operating properly. Thus, they do require some periodic maintenance.

Besides automation, there are other simple measures that hotels use to minimize the amount of energy wasted by staff, guests, and visitors. For example, placing light and equipment control panels in locations that are readily accessible to the workers responsible for operating them, providing indications on larger control panels to clarify the function of each switch and offering instructions on how it should be operated, and fitting guestrooms with a master switch that can be used to easily turn off all lights and fans when guests or housekeepers leave the rooms.

While it is important to save energy by turning off appliances when they are not needed, it is also important to minimize the amount of energy wasted by phantom loads which consume energy while appliances are switched off or in a standby mode. The devices that typically generate these phantom loads include power supplies, battery chargers, transformers (whether or not they are powering any appliances), TVs, VCRs, DVD players, audio systems, computers, printers, copiers, microwave ovens, air conditioning systems, other appliances with remote controls, and devices with stand-by lights or digital displays. Although phantom loads are relatively low, typically ranging from 5 W to 15 W per device, they can account for up to 10 percent of residential electricity consumption and, therefore, can be a heavy burden on facilities that pay a high price for power, whether delivered by the grid or self-generated.

Measures that can be taken to reduce the impact of phantom loads include:

- Unplugging phantom load devices, such as battery chargers, when they are not in use.
- Using a switchable power bar or surge protector with multiple sockets to conveniently turn off devices that are normally used together, such as a computer, a monitor, and a printer.
- Purchasing devices that offer a real "off" switch or have low standby power requirements. A list of "recommended" and "best available" standby power levels for a wide range of electrical devices can be downloaded from the US FEMP's web page (www.eere.energy.gov).

- Avoid using remote controls for equipment that can be operated from hard-wired systems instead (such as air conditioning systems and ceiling fans).

## Case Study

### Ngorongoro Serena Lodge Energy Reduction through Efficiency

Tanzania's 75-room Ngorongoro Serena Lodge, set on the rim of the stunning Ngorongoro Crater, is constructed from local river rock and covered in indigenous vines, blending seamlessly into its environment. The lodge is not connected to the national electricity grid and generates its own electricity with three diesel-powered generators using electronic fuel management systems to maximize fuel efficiency. Trained in-house technicians operate the generators and perform routine maintenance tasks; complex maintenance/repairs are handled by engineers from the firm that supplied the generators.

The generators provide electricity for lighting, ventilation fans, a water treatment plant, laundry and kitchen appliances, a workshop, and a back-up water heating system. The lodge uses efficient lighting and does not employ air conditioning in order to reduce electricity requirements. The lodge meets guest hot water requirements with solar hot water systems. Other facility hot water needs are satisfied with waste heat from the generators' exhaust, a two-stage waste incineration system, and an oil-fired boiler for back-up. To reduce generator fuel/maintenance costs, battery banks are being installed around the facility to enable shut-down of the generators during the day for up to eight hours. Whenever the generators operate, they will use excess generating capacity to charge the batteries. Operating the system this way also reduces noise and air pollution from the generators.

### Example 2. Improving End-Use Equipment Efficiency at Paradise

After conducting the initial assessment, the energy team would make recommendations on how to save energy by replacing existing equipment (behavioral changes will be considered in Step 3). By making several changes, energy usage falls from 50,120 Wh/day to 27,664 Wh/day (savings of 22,456 Wh/day).

The following points highlight the largest opportunities for saving:

- Changing to compact fluorescent lights indoors and LED outdoors from incandescent bulbs—saves 1 1 ,272 Wh/day; a 72% decrease in use.
- Installing efficient fans—saves 4,800 Wh/day; a 50% decrease in use.
- Using an energy efficient refrigerator and freezer—saves 2,784 Wh/day; a 58% decrease in use.
- Using efficient laptop computers—saves 3,000 Wh/day; a 50% decrease in use.
- Installing a flat screen TV—saves 600 Wh/day; a 38% decrease in use.

Although energy efficiency devices will save energy and money over time, the initial purchase may seem expensive. The energy team can look at the cost of making these changes and whether the costs are worth it. Paradise could save 8,196 kilowatt hours a year with energy efficiency equipment (22,456 Wh/day x 365 days = 8,196 kWh/year).

Taking the cost of all the new equipment and subtracting that from annual energy savings, Paradise could achieve a payback period of about eight months. Lighting, due to the comparatively low cost of the equipment and high savings potential, comes out at the top of the list in terms of what to do first; payback for lighting retrofits is often less than a month.

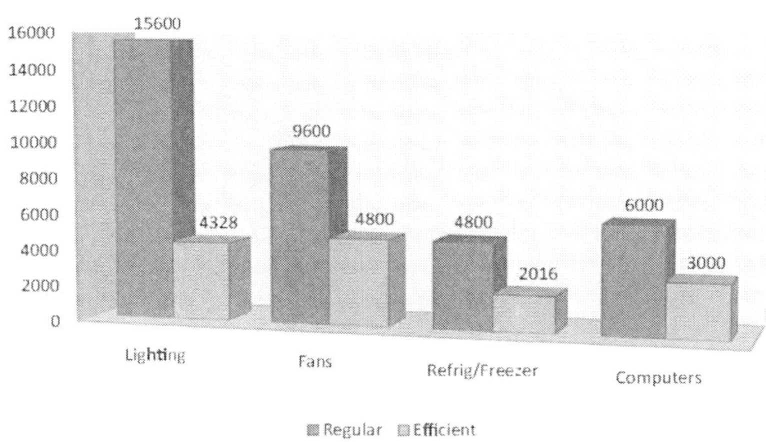

| Item | Quantity | Unit Cost | Total Cost | Wh Saved per Day | kWh Saved per Year | Dollars Saved per Year (kwh saved per year x price per kWh) | Simple Payback Months (total cost/dollars saved per year x 12) |
|---|---|---|---|---|---|---|---|
| Lights in Lodge Rooms | 6 | $5 | $30 | 1824 | 666 | $732 | 0.5 |
| Lights in Bathrooms | 6 | $5 | $30 | 456 | 166 | $183 | 2.0 |
| Fans in Guestrooms | 6 | $100 | $600 | 3600 | 1,314 | $1,445 | 5.0 |
| Refrigerator | 1 | $1,000 | $1,000 | 1392 | 508 | $559 | 21.5 |
| Freezer | 1 | $1,000 | $1,000 | 1392 | 508 | $559 | 21.5 |
| Lights in Kitchen | 4 | $5 | $20 | 1216 | 444 | $488 | 0.5 |
| Lights in Main Lodge | 12 | $5 | $60 | 5616 | 2,050 | $2,255 | 0.3 |
| Guest Computers | 2 | $500 | $1,000 | 1200 | 438 | $482 | 24.9 |
| Lodge Computers | 2 | $800 | $1,600 | 1800 | 657 | $723 | 26.6 |
| TV | 1 | $400 | $400 | 600 | 219 | $241 | 19.9 |
| Fans in Main Lodge | 2 | $100 | $200 | 1200 | 438 | $482 | 5.0 |
| Outside Lighting | 6 | $25 | $150 | 2160 | 788 | $867 | 2.1 |
| Grand Totals | | | $6,090 | 22,456 | 8,196 | $9,016 | 8 |

**Price per kWh is $1. 1 0. The price will vary depending on technology (as shown in Step 5 example)

## STEP 3. MOTIVATE STAFF, GUESTS, AND VISITORS TO CONSERVE ENERGY

Energy efficiency is not just about technology—it is also about human behavior. A tourism facility's energy consumption is affected by how people use electric devices and hot water, and how they adjust indoor air temperatures. For staff, their use of kitchen and laundry appliances, as well as habits in cleaning rooms and performing other tasks, can have a significant impact on a facility's energy consumption. Using more efficient devices and equipment goes a long way toward reducing energy consumption, but without good practices in place to encourage energy efficient behavior, these benefits can be lost. For example, as the table below shows for infrequently used lighting in a bathroom, wasteful use of an efficient light bulb consumes more energy per year (168 Wh/day) than efficient use of a conventional light bulb (60 Wh/day).

Efficient behavior also includes maintaining equipment and operating it properly. Poorly maintained equipment tends to use more electricity and burn more fuel than properly maintained equipment. For example, allowing the heat-exchange coils on a refrigerator to become covered in dust and dirt will cause the compressor to work harder and thereby use more energy to keep the refrigerator's interior cold. Promoting efficient energy use behaviors can be accomplished through staff training and visitor education.

## Working with Staff to Conserve Energy

Workers typically have direct control over 50 to 75 percent of the total amount of energy and water consumed in conventional hotels. Although this ratio is lower in high-efficiency facilities, there is no doubt that employees have the power to make or break a facility's energy and water budgets. As a result, it is essential to ensure the staff is: (1) fully committed to conservation efforts, and (2) able to incorporate efficiency into their daily activities. These two objectives can be achieved through effective staff motivation and incentives, training, and monitoring.

**Impact of Behavior and Efficient Lighting on Energy Use ****

| | | **Inefficient lighting: Incandescent** | **Efficient lighting: Compact fluorescent** |
|---|---|---|---|
| | | **60 Watts** | **7 Watts** |
| Wasteful behavior: Leave light on all day | 24 hours/day | 1,440 Wh/day (60 x 24) (1440 x 365 days x $1.10/kWh) = US$ 578/year | 168 Wh/day (7 x 24) (168 x 365 days x $1.10/kWh) = US$ 67/year |
| Efficient behavior: Turn off light when not in use | 1 hour/day | 60 Wh/day (60 x 1) (60 x 365 days x $1.10/kWh) = US$ 24/year | 7 Wh/day (7 x 1) (7 x 365 days x $1.10/kWh) =US$ 2/year |

**Assuming $1.10/kWh electricity cost and 365 day/year use.

## Motivation

Before learning what they can do to avoid wasting energy and water, workers should know why saving water is important and be motivated to do so. Helping a tourism facility use resources efficiently requires a fair amount of dedication, enthusiasm, and initiative that can only come from workers who clearly understand the reasons for and benefits of conservation.

Motivating workers to participate wholeheartedly in conservation efforts can be difficult for facilities where the principal aim is cost control, but surprisingly simple for properties that embrace sustainable tourism practices. Sustainable tourism earns its name in part because it allows businesses to flourish by operating cost effectively but, more importantly, because they care for the long-term well-being of the environment, the community, and the people who make tourism possible in the first place.

In the long term, staff motivation can be sustained only if driven by a management team that is fully committed to the property's conservation efforts. Management should practice what it preaches and demonstrate its commitment through concrete actions. For example, property managers should:

- Practice energy conservation in their daily activities.
- Regularly track the property's energy performance and set goals/targets for staff. ▫ Discuss energy consumption and efficiency issues during weekly staff meetings.
- Incorporate energy conservation goals/criteria into performance reviews of relevant staff.
- Encourage workers to provide feedback on the property's energy conservation program and new ideas on how to improve energy performance.
- Ensure that energy efficiency is used as a key criterion in all decisions concerning new equipment purchases and property upgrades.

## Incentives

Putting in place incentive programs can be an effective way to improve staff participation in energy conservation and other sustainability initiatives. Although incentive programs need to be tailored to the specific needs and constraints of each facility, ideally they should share the following characteristics:

- Incentive programs must be carefully devised and thoroughly evaluated before they are launched. Backtracking or changing the rules of an incentive scheme in mid-course can negatively affect staff morale and participation.
- The programs must be clearly explained to the staff to ensure there are no misunderstandings over their rules, rewards, and duration.
- The process used to select those who receive the awards and, if applicable, the composition of the awards, must be clear and transparent.

Some incentive programs are relatively simple, such as rewarding top-performing employees or departments at the end of each month, whereas others can be much more complex. An example of a highly effective incentive program is a performance-based scheme that shares part of the financial savings achieved through energy conservation with the whole staff. Appendix 3 contains a sample calculation for a performance-based incentive program.

The nature of the awards given by incentive programs vary from place to place. Although monetary awards often tend to have the greatest appeal, many facilities operate successful incentive programs that offer various types of in-kind awards, gifts, additional days off, paid holidays, and scholarships for employees or their children.

## HOTEL INCENTIVE PROGRAM FROM JAMAICA

(as explained by a retired hotel engineer)

"Soon after starting one of my assignments as Chief Engineer in a large Jamaican hotel, I realized that the performance of my maintenance crew was far from acceptable. It took much too long to complete simple tasks and the quality of the work often was unsatisfactory. This resulted not only in guest complaints, but also wasted substantial water and energy as problems were left unattended for weeks.

Since the main problem was a serious lack of motivation, a simple incentive program for the maintenance team was created. The program consisted in grading each completed job based on quality and difficulty, and, at the end of each month, awarding a prize to the worker who had earned the most points. The reward was a large gift basket filled with products from the hotel's food stores.

The program transformed the maintenance department, virtually overnight, doubling both output and effectiveness. The change in attitude was so complete that workers with time on their hands began collecting extra points by completing jobs assigned to other trades. Accordingly, we ended up with plumbers substituting for painters, electricians for carpenters, etc. This created a few problems, but they were promptly and easily addressed.

Although the monthly prizes were appealing, I am convinced that most of this success came from the fact that the program provided recognition to those who worked hardest and added a bit of fun and friendly competition to the daily grind."

## Training

Employees must be aware that their actions affect efficiency, and that even seemingly minor interventions have an impact on the facility's energy consumption and costs. This idea should be communicated through concrete examples that express the impact of inefficiencies in terms that can be readily understood. Some examples:

- In one facility, turning on the garden lights one hour earlier than necessary increased the generator's diesel consumption by 20 liters per year.
- Letting the three ceiling fans in the restaurant run for one hour longer than necessary at the end of each meal service increased the generator's diesel consumption by 61 liters per year.
- Leaving a range burner needlessly on for one hour per day wasted nine bottles of propane gas per year.

Employees must also be trained regarding how they can help save energy in their daily activities. The training should be as practical and hands-on as possible, define the specific actions employees must take and, when appropriate, explain the reasons for these actions. Department managers often assume that the training and instructions they provide to their staff are clear and complete, but in many cases this is far from true.

Towel reuse programs, which are in place in many facilities around the world but rarely work, provide a clear example of the frequent shortcomings of staff training. In these programs, housekeepers are often asked to change only the towels that guests place on the floor or in the tub, and leave all other towels in the bathroom. These same instructions are generally repeated over and over during the regular housekeeping training sessions to make sure nobody forgets. However, the trainers seldom explain why used towels should be left in guestrooms and why a guest would ever agree to this kind of practice in the first place.

Thus, housekeepers often come to the logical conclusion that this program will only cause them trouble and affect the tips they receive from their guests. As a result, most housekeepers simply ignore the program or, in some cases, leave only the used towels that are sufficiently dry to be folded in accordance with the standard towel presentation format. Given that most facilities do not regularly compare their towel use with occupancy, or gauge guest participation in the program, they can go on for years believing that housekeepers are doing what they are asked and that the program is working effectively.

Whenever possible, training should be reinforced by providing workers with written or pictorial instructions which clearly explain how/why they should carry out these duties and conserve energy and resources in their operations. For example, the following energy and water conservation measures might be included in the housekeeper task descriptions:

- Turn off any lights left on by guests and when exiting guestrooms. Rely on natural light when cleaning, if possible.
- Turn on the ceiling fans only if it is too hot.
- Check for water leaks in toilets, faucets, showerheads, and tub spouts.
- Make sure the toilet flush mechanism works properly.
- Make sure the sink and tub stoppers operate and seal properly.
- Change only the towels that have been left by the guests in the dirty towel basket or designated receptacle.
- Spread the towels the guests want to reuse on the towel racks so they can dry properly.
- Unless soiled, change bed linens only on Mondays and Thursdays.
- Record the number and type of fresh towels and linens used in each serviced room.
- Tie a knot in stained linens so they can easily be identified and processed separately in the laundry.

- Turn off and unplug unused electrical equipment when not in use. For example, even the seemingly small draw of power of a television in sleep mode can use significant amounts of energy.
- Report all identified maintenance problems to the relevant manager.
- Turn off any lights used while cleaning.

As training should not be a one-way street, facilities should also solicit input from their staff on new measures that could be adopted to save energy and other resources. Given that employees are intimately familiar with the workings of their facilities, they are often the best source of information for improving performance.

## Case Study

### Tiamo Resort Working with Staff to Conserve Energy

Tiamo Resort provide luxurious comforts in the wilderness setting of South Andros Island, Bahamas, the largest and most undeveloped of the many islands of the Bahamas archipelago. This high-end hotel uses solar electricity and has no air conditioning, no internet connection, no televisions, and only a single public telephone. Its elegant open-air beachfront bungalows are cooled by natural breezes and ceiling fans, water is heated by the sun, and bathrooms are fitted with high-efficiency composting toilets which consume only half a liter of water per flush (12 times less than a standard water-saving toilet).

The resort has installed a solar photovoltaic (PV) system that is the largest in the Caribbean for a single resort facility. The 120 kW system is composed of nine arrays in a half-acre field. The systems feed 88 6V lead acid batteries with 4,000 amp-hour battery storage capacity. Tiamo also has engine-generator sets to keep the batteries topped up, which will prolong the battery lifetime. The generators run on a biodiesel blend made from cooking oil collected from cruise ships.

Working with the staff to examine energy usage and behavior for daily operations, Tiamo management implemented conservation measures to reduce energy consumption. Paying attention to energy consumption, monitoring, and focusing on improved efficiency has led to large and small decisions about energy usage. These have included installing meters to monitor energy consumption, using French presses rather than electric coffee pots to make coffee, using clothes lines to dry clothes rather than electric dryers, and using grey water to irrigate resort vegetation. Producing 100 percent of its energy from alternative sources and utilizing sustainable principles has made Tiamo one of the most environmentally-friendly resorts in the world.

The remote location of Tiamo Resort makes energy independence extremely important. The photograph shows solar panels at Tiamo used to generate electricity for a range of applications at the resort.

## Visitor Education

For facilities with a sustainability theme, such as eco-lodges, nature-oriented hotels, or park facilities, visitor education programs about the facility's sustainable energy practices can be a positive addition to the overall visitor experience. When appropriate, visitor education materials can explain the environmental benefits of the facility's energy practices. Benefits can range from avoided air pollution and noise from generators, to reduced waste and consumption of fossil fuels.

Because visitors come from varying parts of the world, they may have different ideas about comfort, including room temperature, bath water temperature, length of showers, and opening and closing of windows. All of these practices will impact the facility's energy consumption. Visitors from different countries may also be accustomed to different kinds of electrical fixtures, heating and cooling equipment, and electrical devices. For these reasons, visitor education should cover topics that may often seem obvious, such as the location of electrical switches, operation of in-room heating, and conscious use of hot water.

Visitor education should have the goal of providing information about ways to minimize energy use during guests' stay without sacrificing comfort. Educating guests on what the facility is doing to conserve energy showcases the sustainability of the operation. Options to communicate the facility's sustainability efforts include:

- Tours highlighting the facility's renewable energy equipment and explanations of its energy conservation measures. By providing insight on sustainability and what it takes to run an off- grid facility, well-crafted tours can be a valuable addition to the guests' experience.
- Informational booklets explaining how to conserve energy.
- Explanation of conservation measures by facility staff at check-in.
- Reminder notices posted in rooms and around the facility.
- Information on the hotel's website (good examples can be found at www.hotelmockingbirdhill.com and www.tiamoresorts.com).

Guests should be invited and motivated to help the facility save energy. However, it should be recognized that many guests may feel they are paying enough and should not be asked to sacrifice their comfort to help the facility save money, regardless of difficulties in obtaining fuel and producing electricity. Asking for and winning the guests' cooperation is significantly easier for facilities that embrace a broader sustainability approach, and show as much concern for the well-being of the environment and host communities as for the efficiency with which they use energy and other resources. Guests can be encouraged to do the following:

- Turn off guestroom lights and appliances when they are not needed and before leaving the rooms.
- Open and close window shades depending on the time of day.
- Use towels and linens for more than one application.
- Dispose recyclable wastes in the facility's recycling bins.
- Take home batteries and other hazardous wastes they generate, as these items cannot be properly disposed of at the site.
- Schedule activities around availability of energy resources, such as showering in the evening instead of the morning in instances where solar hot water is in use.

## Example 3. Encouraging Staff and Guests at Paradise Resort to use Less Energy

In Example 2, the daily electricity requirement was reduced by switching to more efficient appliances—especially lighting and fans. By training staff to turn off lighting in the office area when it is unoccupied, lighting usage can be reduced even further (especially at night). Staff members and guests can shut down computers and turn off monitors when not in use. Camera/cell phone battery chargers should be unplugged when not charging. To ensure these measures are followed, a reminder notice should be posted in the office area by the light switch and on the computers. Staff can monitor outlet use in the lodge. Guests can be encouraged to use non-electric razors and let hair air-dry instead of using hairdryers and electric razors (items that consume a large amount of electricity). By working with staff and guests to change behavior, Paradise can save 2,510 Wh/day, a further 9 percent reduction. After implementing all energy efficiency measures, Paradise's daily energy demand drops to 25,154 Wh/day.

# STEP 4. ACCOUNT FOR FUTURE CHANGES IN ENERGY NEEDS

Design of an energy system for an off-grid tourism facility must consider future energy demand as well as immediate needs. Expansion plans should address strategies for either: (1) improvement in energy efficiency to reduce the energy per guest requirement; or (2) investment in new equipment to increase on-site power generation capacity.

Quantifying a facility's energy requirement in energy units per room or per visitor makes it possible to estimate the impact of facility growth on energy demand. When the facility's long-term plan includes an increase in the number of rooms or visitors, that increase can be included in estimates of future energy requirements. For example, lodging facilities can calculate the average daily energy consumption per guest room in kWh/room-day. Facilities that do not provide lodging services may quantify energy use in kWh/visitor-day. Energy service professionals can develop more detailed estimates and projections that account for seasonal variation in energy demand.

Changes in the types of services offered at a facility will potentially impact future energy needs. Sizing the energy system up front to handle extra capacity, or designing the system to add extra capacity relatively easily, can save time and money in the long run. When sizing the system, it is appropriate to add a certain percentage (usually 10 percent) energy demand factor to a daily watt-hour summary to account for "load creep" and errors in judgment for hours of usage, but systems should not be sized based on expected loads many years in the future. For example, if a facility installed solar power, the system should be designed to ensure the equipment room is large enough to add inverters and batteries, and that there is adequate space to install additional solar arrays. In this way, the system can be expanded when extra power is needed and new capacity added relatively quickly and easily.

## Example 4. Growth Plan For Paradise

In the original site/facility assessment, the assessor would ask the owners many questions to determine where they hope go with the facility in the next 5 to 20 years. It is important to know these estimates when designing the energy system. Depending on the type of system used, the designs for future growth will differ (see Step 5 in main text).

If the owners would like to double the number of guest rooms in the next five years (keeping the lodge amenities the same), Wh/day energy consumption likely would increase about a third from 25,154 to 36,000. The main energy increase would be from additional lighting and outlets in the guest rooms.

Depending on the energy system in place, different allowances for future energy use should be made. For example, if the facility is using a solar PV system, extra physical space should be saved for the expansion of the solar array, and the installation of battery and inverter equipment. It would not be appropriate to install upfront a PV system that is twice as large as what is needed now, as the entire system most likely would be too expensive. For a generator, it is wiser to purchase an additional unit as the load increases rather than initially purchase a larger machine that can accommodate potential future needs.

## STEP 5. EVALUATE ON-SITE ENERGY SUPPLY OPTIONS

Now that we have examined the factors that influence energy demand and consumption, it is time to address energy supply. From a technical perspective, evaluating technologies for on-site energy supply requires answering two questions:

1. What energy resources are available at the facility site?
2. What energy supply equipment can meet the facility's energy needs given the available resource?

The choice of an on-site energy supply system will vary from facility to facility. An energy supply system that is technically feasible may not be appropriate for other reasons, such as unavailability of maintenance and repair services, local regulations prohibiting certain types of equipment, or high cost of equipment or fuel. Some of these issues will be addressed in Steps 6 and 7. Additional questions to ask when evaluating options for an energy supply system include:

- What kinds of equipment can local suppliers sell, install, and service?
- What kind of financing is available?

- What is the cost and degree of technical expertise required to operate and maintain a given energy supply system?
- What is the environmental impact of the system?

## On-Site Electricity Generation Options

Energy resources for on-site energy generation include fossil fuels, solar radiation for photovoltaic systems, wind, hydropower, and biomass. The following table summarizes the key features of each option, highlighting cost per unit of energy, life of the system, operation and maintenance concerns, and potential environmental impacts. The most common technologies are described in more detail below.

**Engine-Generator Sets** are the most common type of energy generation found in rural areas. They often run on fossil fuels, especially diesel, but can also run on gasoline, propane, and biofuel. Engine-generator sets typically are rated according to the electric output of the generator in kilowatts, and system designers choose a generator size that meets the peak electric demand plus a safety margin of generally 20 percent. Engine-generator sets run best when loaded at over 50 percent of their rating, and obtain their best fuel efficiency when loaded at 50 to 80 percent of their ratings. During periods of low electric demand, engines consume more fuel per unit of electrical output and experience additional wear and tear on the engine. In some cases, using multiple smaller engine-generator sets instead of a single large one may improve efficiency and reduce maintenance requirements. Diesel engine-generator set lifetimes depend on how the systems are operated, and the level of maintenance.

A well-maintained diesel generator will typically last about 15,000 operating hours before it needs replacement or a major overhaul. Periodic maintenance includes regular engine oil changes and minor overhauls.

Generators usually have the lowest up-front cost compared to other power generation options. However, they do have drawbacks. If the generator is run on fossil fuels, reliability of supply, fuel cost, price volatility, and environmental impact of storage and use must be considered. Generators run on fossil fuels produce greenhouse gas emissions, exhaust fumes, and are noisy. Even when used efficiently, they may leave a considerable environmental footprint.

# CASE STUDY

## Posadas Amazonas Using a Single Generator for Basic Energy Needs

Located on the edge of an old growth tropical forest in Peru, Posadas Amazonas is a community-based tourism venture. Rooms in the lodge open directly onto the neighboring 1.8 million acre Tambopata National Wildlife Reserve. There is no grid electricity or hot water at Posadas Amazonas and all biodegradable material is composted on-site. A single 5,500W gasoline-powered generator provides electricity for the kitchen appliances and for common areas. The generator also provides back-up energy for the office's small photovoltaic system, which powers a radio and computer. The gasoline used to run the generator is readily available and is also used to power boats, which are the main means of transportation to and from the lodge. The generator runs about five hours per day, when the guests are away on excursions into the rainforest. The generator consumes about 375 liters of gasoline per month. Kerosene lamps illuminate hallways, common areas, and guest rooms. Candles on tables and candelabras light the lobby and dining area. The facility uses natural gas for cooking.

A kerosene lamp is used for illumination at Posadas Amazonas

**Solar Energy** can be used to generate electricity or heat water. The most common use of solar power for electricity generation is in photovoltaic systems which use solar panels to collect sunlight and convert it to electricity. PV systems are excellent for handling small and moderately sized loads and offer the advantage of being modular—making it possible to add capacity as electricity demand grows. These systems are quiet and produce no emissions, making them a good choice for environmentally sustainable operations. Typical PV systems for electricity generation include a battery bank that can store the energy produced by the solar panels, a charge controller to ensure the battery is not over charged or over discharged, and an inverter to convert the direct current (DC) power produced by the PV panel to alternating current (AC) power required by most electrical devices.

Well-maintained PV systems can last 25 years or more, but will require periodic replacement of inverters (5 – 10 years) and batteries (3 – 10 years depending on use and maintenance). PV systems can be one of the most expensive up-front energy options, but once the system is purchased, may be economical over the long-run. Additional uses for solar energy include smaller independent devices such as solar-powered flashlights, path lighting, and lanterns.

The amount of solar energy that reaches a site depends on the site's latitude, seasonal weather patterns, and availability of open, unshaded space for solar panels. Estimates of solar resources are available for any location on earth from the Internet (see References section for a listing of renewable resource websites).

**Wind Energy** can be used for on-site electricity generation using turbines to convert wind to electricity. Wind turbines appropriate for off-grid energy supply systems range in size from 500 W for battery charging to about 100 kW for use in hybrid systems that combine wind turbines with an engine-generator set or PV panels. In the most commonly used type of wind turbine, a rotor consisting of two or three propeller-like blades captures the energy in the wind and spins a horizontal shaft connected to an electric generator. Wind turbines require periodic inspections and preventative maintenance. Turbine lifetimes depend on the turbine's design, but a properly maintained turbine should last 10 – 15 years. Like solar power, wind energy does not produce emissions and is considered a carbon-neutral technology.

The wind resource at a particular location depends on seasonal wind patterns in the region, the local geography, and exposure of the site to wind. The site suitability for wind turbines is highly dependent on the specific conditions at the site, including buildings and trees that may block the wind. Wind resource professionals can make very rough estimates of wind conditions at a site based on a site visit and meteorological measurements. For any large wind installation, a site study should be completed, measuring the wind strength, 24 hours/day for at least a year.

**Hydropower** systems use water power to generate electricity or provide power for such applications as water pumping or grain grinding. Some hydropower systems generate waste heat that can be used for heating applications. Other systems use batteries to capture the energy. For all hydropower applications, the hydro resource is a measure of the energy available in flowing water as it moves from higher to lower elevations. This is a function of the distance the water falls through, and the quantity of water that falls. Some hydropower systems produce electricity 24 hours per day, so even a small system can produce large amounts of energy. Unlike other energy supply options, hydropower systems are not modular and require a good deal of infrastructure to set up. The maximum size of a system is fixed by the available hydro resource. A well-maintained small hydropower system can produce power for 30 years or even longer.

It is essential that hydropower systems be carefully designed to create the smallest impact on the environment. Although they produce no emissions, hydropower systems can negatively affect fish migrations and river banks and cause erosion if not situated properly. Additionally, hydropower can be seasonal—if there is a drought or low rainfall, the system can be out of commission for months at a time.

**Biomass energy** can be used in gasifiers for electricity generation, or burned directly for heating applications. Biomass resources can include wood, agricultural residues, biodegradable waste, and crops raised specifically for energy. Gasifier systems used for electric power generation typically convert biomass to electricity at the rate of about 1 – 3 kg of dried wood chips per kWh of electricity. These gasifiers are used in electric power systems rated between 10 kW and 50 kW. Biomass gasification systems have high operation and maintenance requirements, and require a carefully managed sustainable supply of biomass fuel.

Because resources vary widely by location, it is impossible to make generalizations about the biomass resource availability at a particular site. Factors to consider in the assessment of the biomass resource include sustainability of the supply, seasonal variation, and cost of collection, transportation, and processing. The biomass resource for small power applications is often measured in tons per month.

**Hybrid systems** utilize more than one of the energy resources described above in a single system. Often hybrid systems combine an engine-generator set with solar or wind power. These systems are often an excellent solution to meet energy needs as they can utilize the positive aspects of two or more types of generation options. For example, solar panels are quiet, produce no emissions, and require no fuel. However, they need batteries to store energy and can be inoperable during cloudy days. Coupled with a generator for back-up power, solar panels can be a viable option for many sites. Hybrid systems must be carefully designed and operated to ensure the equipment is working well together. This requires on-site technical expertise and will require close cooperation between the system designer and the operator, as well as the operator and equipment provider.

## Energy Storage and Conversion

Although some energy generation options can directly run devices, others require electricity conversion or storage for use when demand increases. Electricity from the grid comes in the form of alternating current (AC) power which is used by most commercially available electrical appliances, lighting fixtures, and devices designed to be plugged in to a wall socket. Direct current (DC) is the type of electricity stored by batteries and some types of electric generators, and is used by battery-operated devices and special DC light fixtures and appliances. Photovoltaic panels produce DC electricity, which is stored in batteries. In some small systems, this DC electricity can be used directly by DC lighting and special DC appliances. Generally, PV systems include inverters that change this DC electricity to AC electricity for more general consumption.

**Batteries** can be charged when power is in abundance, and can be used when power generation systems or sources are unavailable. For example, in PV systems, batteries store power during the day when the sun is strong, and make it available at night or during cloudy periods. Batteries can be used with engine-generator sets to improve system efficiency; the generator runs at full power to meet both the electric demand and to charge batteries, and then shuts down when demand is low, allowing the batteries to deliver power. In addition to solar and engine generator sets, batteries can also be used with wind and hydropower.

**Rural Tourism Facility Power Generation Options.**

| Technology | General Description | Initial/ Capital Cost** | Operating Cost ($/kWh) | Expected Life | Operation and Maintenance | Environmental Impacts |
|---|---|---|---|---|---|---|
| Engine-Generator Sets | Lowest initial cost of systems; flexible sizing; versatile fuels; perform best when loaded more than 50 percent. Require regular maintenance; trained staff; often installed in pairs. | Initially low at $0.50 to $1.00 / watt | Operating costs are very high $.50 to $2.00/kWh not including initial costs (Same range including initial costs) | 15,000 hours 1.7 years operating 24 hours/ day 5 years operating 8 hours/day | Requires regular maintenance and lubrication fluids. On-site mechanic may be needed, as mainten-ance is critical. Minor overhauls and complete overhauls are schedu-led expenses; fuel is delivered in 55 gallon drums; quality of fuel may be a problem in some places. | Noisy; air pollution; waste oil disposal or return, risk of spills. Disposal/ recycling is rare—abandoned engines are typical. Carbon emissions from fossil fuels. |
| Photovoltaic system | Highest initial cost. System includes PV panels, batteries, inverter, charge controllers, and load controlling equipment. | $10 – $1 8/ kW (of solar panels) | $0.25/kWh not including initial costs $1.00 to $1.25 / kWh including initial costs | System: 20 years. Battery life 3 – 10 years Inverter 5 –15 years | Require a savings program for battery replacement and expansion fund: $20 per month per battery. Inverter may require replacement or upgrade every 7 to 10 years. Need monitoring system for performance. | Lead acid batteries should be recycled or chemicals may leach into the ground. No carbon emissions. No noise. |

**Rural Tourism Facility Power Generation Options (Continued).**

| Technology | General Description | Initial/ Capital Cost** | Operating Cost ($/kWh) | Expected Life | Operation and Maintenance | Environmental Impacts |
|---|---|---|---|---|---|---|
| Hydro Power Energy Systems | Low initial cost. Takes falling water and converts to electricity with a micro-hydro turbine, a generator, a load controller, and distribution wiring. | $2 – $7 per watt depending on sizes | $0.25 to $0.50/ kWh including initial costs. $0.10 to $0.20 not including initial costs | 5 – 15 years depending on technology and location | Regular maintenance is required at the intake to keep it clear of debris, and regular maintenance items are required on the turbine, generator, and distribution wiring. | Water rights are significant issues that must be addressed. Protection for fish and ecosystems is required. Planning for extreme weather event preparation is critical. |
| Wind Energy systems, Stand-alone or hybrids | Less costly than PV but greater maintenance. Still use battery systems for storage of energy. Turbines most suitable for eco-tourism are in the 500 to 5000 watt range. | Small $3.50 – $6/ watt | In the range of $0.50/kWh including initial costs $0.15 to $030/kWh not including initial costs | 10 years; with good service can get 20 years Batteries 3 – 10 years depending on depth of discharge Hybrids with PV and wind work well | Wind systems require wind resource. Proper site analysis is needed to locate turbines and prepare energy production estimates. The larger the project the more detail to siting is needed. Installation logistics are reduced for small systems. | Site location is a major issue for large and medium turbines due to tower height and visibility. No emissions. |

| Technology | General Description | Initial/ Capital Cost** | Operating Cost ($/kWh) | Expected Life | Operation and Maintenance | Environmental Impacts |
|---|---|---|---|---|---|---|
| Biomass Power Generation Systems | Suitable for use with smaller renewable power systems; can generate electric power or produce methane gas for combustion. Feedstocks include wood or agricultural byproducts | $.80 – $5/ watt | Cost of feedstock (depending on availability and transport) | 10 – 20 years | May require backup power. Source of biomass product must be readily available. Requires regular maintenance. | Can eliminate waste stream— including animal and even human waste—through use of a digester. Composting can present odors. Potential impact of sustainable feedstock collection and gasification (emissions). |
| Solar powered water pumps | Solar water pump system design depends on the quantity of water required and the height to which it must be pumped. An elevated water storage tank can ensure good water pressure throughout the facility; water conservation is also important. | $20,000 to $50,000 | Minimal, as there are no batteries | Pumps 5 – 10 years; replacement parts needed | Pumps are vulnerable to dirty water and salty environments. Pump life depends on usage, but 5 – 8 years is not uncommon. Water fixtures should conserve water. | Source of emission-free water. |

**Costs are labeled "initial/capital cost" and "operating cost." Initial costs cover the purchase and installation of equipment. Operating costs include fuel, battery replacement, parts replacement, labor, etc. that are incurred during use of the equipment. Note that these are costs per installed watt of equipment and not a measure of the amount of energy the equipment will produce. For example, with solar power, initial costs quoted are $/watt of the solar panels. With hydropower, the costs quoted are generally $/watt of the turbine/generator. However, a 1,000 watt solar panel array may be able to produce 3,000 watt hours of energy/day. A 1,000 watt micro-hydro turbine can produce five or six times as much energy, because it produces energy even when the sun is not shining, as long as the water is running. Costs are derived from the HOMER model (www.nrel.gov) and may vary from actual costs in any given location.

## Battery Maintenance Inspection Lists for Solar Array

### Weekly Maintenance Inspection:

1. Measure battery voltages in the morning, noon, and night at the same time each week.
2. Record array voltage and, if available, currents from solar array to battery.
3. Ensure modules are clean and free from dust and shade when measurements are made.

### Monthly Maintenance Inspection

1. Check water volumes in flooded batteries.
2. Inspect battery connectors so they are free of corrosion and not loose.
3. Ensure that system is performing to its capacity.
4. Trace battery voltage for the four weeks.
5. In large battery banks, equalize the battery array by charging all batteries to achieve equal voltages.

### Annual Inspection

1. Check connectors on alternating currents.
2. Conduct current testing to determine amps flowing from modules to batteries and amps flowing to inverter under a constant load.
3. Test battery capacity with hydrometer or by load testing.

Battery lifetimes are very sensitive to their operating conditions. They must be prevented from being too deeply discharged or from being overcharged. Batteries like to operate in a partial state of charge. The battery life span will be greatly extended if the battery is operated at more than 50 percent of full charge. Discharging the battery below 40 percent will reduce the number of cycles the battery will perform. Batteries must be kept clean and free from corrosion on the terminals. Hybrid systems with a back-up engine allow a system to bring batteries back to a comfortable battery state of charge. Operation and maintenance plans for systems that include batteries should include funds for maintaining batteries and for replacing them approximately every three to five years (the actual number will vary with the use and maintenance of the battery). Proper battery use and maintenance can prove challenging for many new system operators, especially those utilizing PV.

**Charge controllers** limit the rate that the electric current is added to the batteries. They prevent overcharging and over-draining the batteries to both increase the lifespan and performance and reduce the risk of overvoltage.

**Inverters** change DC electricity from PV panels and batteries to AC electricity needed by many electric appliances. If used properly, inverters require little maintenance and have lifetimes of between five and seven years. However, inverters are often set incorrectly, resulting in problems. It is essential to work with trained operations and maintenance staff to ensure correct operation.

## Options for on-Site Heating Applications

On-site heating applications are used for cooking and baking, heating water for bathing and washing, and space heating where necessary. The following table provides a comparison of heating application equipment. Unlike power generation options, which all provide the same output—electricity, heating applications utilize different types of fuels and generate different outputs, making cost comparisons more complex as there is not one unifying measurement.

**Solar water heating systems** collect heat from the sun to heat and store hot water. Solar water heating systems differ depending on the climate in which the system is located. Warm climates can use simpler designs that store solar-heated water during the day for use at night or the next morning. Colder climates circulate antifreeze and transfer the heat to water, storing it in tanks each day when the sun is available.

**On-demand water heating systems** heat water as it is needed. For example, a bathroom on- demand system will turn on when someone opens a bathroom hot water faucet. Propane powered on- demand systems are reliable but may have high operating costs and typically need high pressure gas. Electric systems which are the norm in many parts of the world are less reliable and more dangerous but may be appropriate for facilities with a single hot water pipe. Converting gas propane units to biogas should be feasible with some adjustment to gas orifices that need to supply more fuel than propane.

**Solar hot air systems** can circulate interior or exterior air and deliver it back to the building after being warmed in a solar collector. A vertical air panel can heat air directly without a fan. By adding a PV-driven fan, greater air flow can be achieved and more energy can be collected. The air moving turbulently through the collector is more efficient. Solar hot air can be used for drying food as well as warming rooms. These systems are most cost-effective when used in cold climates with good solar resources.

**Biogas digesters** use an anaerobic chemical process to convert agricultural, animal, human, or market waste into methane gas that can be burned in gas-powered cook stoves and water heaters. Some types of waste may require processing before they can be introduced into the digester. Although digesters have higher upfront costs than some other options (parts and labor for initial construction), once installed, they tend to be very low cost. Operations and maintenance requirements are low— both in time and skill required.

**Propane gas burners** for cooking can heat evenly and quickly as do modern kitchen appliances that use propane. The fuel is bottled and transported in cylinders. However, propane costs can be high, particularly when the fuel needs to be transported over a long distance.

**Solar box and concentrating cookers** use mirrors or other reflective material to focus solar radiation onto a dark-colored cooking pot enclosed in a glass-covered case. Solar cooking may be a viable option for tourism facilities, provided a back-up cooking option is available. Solar cooking typically is slower than gas/biomass and some cookers need to be constantly tended to make sure they are properly angled to catch the sun's rays. In areas that have staff around the kitchen all the time (as well as adequate solar radiation), solar cooking can be a viable option.

**Cooking and Heating Options for Off-Grid Facilities.**

| Technology | General Description | Cost | Expected Life | Operation and Maintenance | Environmental Impacts |
|---|---|---|---|---|---|
| Solar water heating systems | Provide hot water for bathing, for kitchen, and for heating systems. Can also provide cooling for commercial systems. | For water heating: $4,000 – $10,000<br>For heating system: $10,000 – $20,000 | 10 – 20 years depending on O&M | Solar heating systems are generally maintenance- free and in some cases have no moving parts; some include pumps and controls that are subject to electrical conditions. Antifreeze may require replacement every 5 – 8 years depending on operating conditions. | No emissions or noise pollution; sustainable option. |
| On-demand water heating systems | Units are limited in size and are available in gas and electric units. Electric shower heads are very dangerous as they mix water and high current electricity and use a lot of power for demand hot water. Gas units are common but require propane—at high pressure could operate on biogas. | Electric shower head heaters: $20 – $100<br>Propane on demand: $300 – $2,000 depending on make, model and country | 3 – 5 years | On-demand heater only reduces storage of heat. System uses high energy to heat incoming cold water Water quality is an issue with all water heaters as minerals will settle when units cool down. Groundwater needs testing to be sure it is not acidic, which reduces the life of these units. | The high energy requirement from electric on-demand heaters can cause high energy demand. Propane needs a tank for storage; emits greenhouse gases. |

**Cooking and Heating Options for Off-Grid Facilities (Continued).**

| Technology | General Description | Cost | Expected Life | Operation and Maintenance | Environmental Impacts |
|---|---|---|---|---|---|
| Solar hot air systems | Brings fresh air into a home or room and pre-heats it through the collector—can use a control-operated blower to circulate air when the sun's heat is available. Hot air heaters are useful in colder climates but without storage they last only during the peak hours of sunlight in the winter. | Hot air direct, no storage: $5000 – $8000; Storage adds $5000 Cost effective against fuel heating systems | 10 – 25 years Blower and controls may last only 10 years | Filter requires changing; if storage added, requires periodic maintenance. | No issues. |
| Biogas digester for cooking | Most complex type requires filling a batch-type digester tank with waste and/or biomass to right proportions and creating anaerobic digestion and collection reservoir for gas. | Highest cost of cooking options due to the need for digester and reservoir; $4,000 – $20,000 depending on location, size and type of digester | 25,000 hours or 3 – 5 yrs | Requires regular maintenance and refilling every 6 – 8 months depending on materials used. Waste stream of material is needed to keep active culture alive. Loading digester can be simplified but is a chore. Appliances may need to be modified to burn biogas vs. propane or methane. | Gas is produced free of odor so some danger exists in detecting it. Also, often hard to see flames. Sludge left after the digestion can be used to make compost for gardening. |

| Technology | General Description | Cost | Expected Life | Operation and Maintenance | Environmental Impacts |
|---|---|---|---|---|---|
| Propane gas for heating water and for cooking | Uses conventional appliances and propane. | Fuel costs parallel oil costs | 10 – 20 years | Little maintenance; subject to fuel price fluctuations. Extremely durable and available worldwide. | Requires a tank for storage; the large tanks have aesthetics issues; emits greenhouse gases. |
| Gasification stoves | Gasifies materials to produce combustion, less smoke, more heat control. | $5 – $1 00; Vary widely | 3 – 5 years | Heating wood chips or organic matter releases gases that burn, creating heat. | Burns with less smoke once ignited. |
| Solar ovens | Well suited to baking—highest temperatures 300 degrees, without reflectors. Typically constructed of local materials. | Imported units can cost from $100 – $400; costs can be higher depending on size | 1 – 2 years for cardboard | Cooking time may be slow and oven may need periodic adjustment to track the sun. Reflectors must be kept clean. Glazing is also important and must be transparent. | Little impact if built with local materials. Benefit of not cutting trees for fire wood. |
| Solar reflector | Good for high temperature short-term cooking—frying and boiling water. Risk to eyes if not used carefully. | $10 – $300 depending on type of product | 1 – 5 years | Panels must be kept clean. Panels can be blown away/knocked over in very windy locations. | No issues. |
| Efficient charcoal and wood stoves | Utilize traditional cooking methods and materials but should cook faster and with less smoke and fuel consumption. | Various cost $5 – $50 | 2 – 5 year. | Most stoves require little maintenance, but chimneys must be kept clean. Safer and cleaner than open fires. | Less polluting than open fire; greenhouse gasses emitted during combustion; feedstock sustainability. |

**Wood and charcoal stoves** can be much more efficient than open cooking fires. A good stove, coupled with a well-trained cook who knows how to manage the fuel source, can minimize fuel consumption. Stoves utilizing ethanol and plant-derived oils instead of biomass are also now available in some countries.

## Water Pumping

Water pumps can be driven electrically or mechanically—electric water pumps use any of the electric power options and mechanical water pumps can be driven by engines, water power systems, or mechanical wind pumps. Animal and human-powered water pumps are also available. Water pumping systems typically include a water storage tank, which makes water pumping well-suited for use with wind and solar power. Based on an assessment of the solar or wind resource, and the facility's water use requirements, the system can be designed so the pump operates whenever the sun or wind is available, and fills the storage tank to meet the water demand. The cost of these units ranges from US $20,000 – US $50,000 and they can last from 5 – 10 years (with replacement parts needed). Energy required for water pumping is determined by the vertical upward distance that the water must be raised, the horizontal distance that it needs to travel, and the daily pumping requirement.

## Example 5. Options and Costs for the Paradise Energy System

Based on the data gathered in the previous steps, Paradise Resort will require an energy system capable of producing approximately 25,000 Wh/day (or 25 kWh/day) at present with a potential of 36 kWh/day in the next five years (based on its business growth plan). Below are the power generation options that can be considered for Paradise: solar, solar hybrid (with a diesel generator), diesel generator, micro-hydro turbine, and wind. As shown below, the costs vary widely depending on technology, but this is the basis for the estimate of $1.10 kWh used in the previous examples.

**Solar Alone:** The solar array is sized at 12 kW to be more reliable, since there is no generator. It is estimated that this will not meet load requirements about 2 percent of the time due to cloudy days, weather, etc. (this is a generally acceptable percentage for solar).

| | Initial Capital Cost | Operating Cost $/Year | NPC (Full system cost over 20 years) | Cost of Energy $/kwhr |
|---|---|---|---|---|
| Solar | $152,000 | $1,872 | $182,616 | $1.109 |
| Solar/Diesel Hybrid | $137,000 | $2,600 | $179,520 | $1 .074 |
| Generator with Battery/Inverter | $27,667 | $10,446 | $198,477 | $1. 1 88 |
| Generator Alone | $19,000 | $15,942 | $279,674 | $1.674 |
| Micro-Hydro | $39,000 | $1,200 | $58,622 | $0.35 1 |
| Wind | $48,000 | $1,908 | $79,205 | $0.490 |

** These costs are derived from the HOMER program and could change with variations in the cost of fuel, site conditions, and the financial interest rate. The simulation is run with the cost of fuel at $1.50/liter and the interest rate at about 2 percent. Rises in fuel costs will have a strong impact on the net present cost (NPC) of the generator-only option. Rises in the interest rate tend to make high capital-cost projects less attractive.

**Solar/Diesel Hybrid:** This system includes a 10 kW PV system with a 5 kW generator. The generator is included to cover the times when it is cloudy, or the batteries need a boost. It enables the solar system to be sized a little smaller and is a good safety precaution. In this case, the generator only burns 356 liters of fuel/year.

**Generator/Battery/Inverter:** A generator is connected to a battery inverter system to allow the generator to be turned off periodically. When the generator runs, it charges the system batteries. The initial cost is lower than some other options, but it is subject to fluctuations in the price of oil, and has issues of pollution and noise.

**Generator:** A generator alone has the lowest capital cost of all options, but the highest operating cost and the highest cost if looked at all technologies over 20 years. Like the option above, price, pollution, and noise are issues.

**Micro-Hydro System:** If a suitable water source is available, a micro-hydro system can be the cheapest form of electricity. If Paradise were close enough to a water resource that would permit the installation of a 5 kW micro-hydro turbine, this would be the cheapest option. The water would need to come from a small to medium stream with a lot of fall, at maximum 0.5 kilometers from **the facility.**

**Wind** System: The wind system would include two turbines in a relatively good wind environment, generating DC power (like a solar PV system) with controller and battery equipment similar to a PV system.

## STEP 6: OBTAIN FINANCING, PERMITS, AND EQUIPMENT

Implementation of an on-site energy supply project depends on the conditions at the facility's location. Access to capital and financial institutions, proximity to equipment suppliers and service providers, and national and local regulations will all influence how a project is implemented. In some areas, regulations may promote tourism business development by providing financial incentives, and energy service providers may provide a complete package of lending and installation services. In other areas, the tourism facility may be on its own to find financing partners and installation service providers, as well as to learn about and comply with applicable regulations.

### Procuring and Installing a Small Power System

Whenever possible, energy efficient and power generation equipment should be purchased from local suppliers who can honor manufacturer warranties and provide maintenance and repair services throughout the life of the project. Suppliers that are not local will often have higher costs for transportation, communication, training, and after-sales service. Depending on the location, some technologies may be more readily available than others—a factor which may influence the decision of one technology over another. Additionally, there may be existing local/regional supply chains, infrastructure, technical capacities, and credit available for certain technologies.

## PROCUREMENT RESOURCES

The World Bank has developed a free online informational toolkit (the RETоolkit) for renewable energy in developing countries. The "Stand

Alone" system section of the toolkit can assist rural facilities with the following:

- Identify and design feasible projects.
- Determine appropriate policies and regulatory frameworks.
- Identify business models.
- Evaluate financing mechanisms.
- Select appropriate and feasible technologies.
- Utilize project tools, including technical standards, generic terms of reference, and supporting documents.

For more information, see the REToolKit: http://go.worldbank.org/ZJNBLHEO50

The USAID Energy Team has also developed draft procurement guidelines and specifications for generators and solar systems being procured for health facilities that can serve as a good guide. http://www.poweringhealth.org/index.shtml

A good way to see what systems are available is to contact local vendors, industry associations, and even utilities. For example, in Kenya and Zimbabwe, existing rural businesses and networks (such as hardware stores and battery companies) facilitated the expansion of small solar systems. In Sri Lanka and Bangladesh, a rural micro-finance institution partnering with a manufacturer helped to create a market for solar systems. For more information and other case studies, refer to the World Bank's REToolKit (see description to the right). A final benefit of working with a local supplier is not having to import a system and arrange for any customs duties or international transport.

When procuring an energy system, it is important that the vendor or your energy consultant properly size the system based on both a resource assessment and an energy audit (a simplified form of which has been presented in the first five steps of this guide).

The services of a qualified energy consultant can be very valuable, as she/he can conduct the audit, write system specs for potential vendors, and review bids from different parties. It is important that the system be installed by qualified professionals to avoid potential system failures. Using new, high-quality equipment with international quality labels and manufacturer warranties will also help ensure that the system operates as designed. Typical supplier warranties would include: (a) honoring all manufacturers' warranties, which would generally be for two years or more on inverters and charge controllers for PV systems, and (b) a one- or two-year warranty on the labor and miscellaneous equipment (one year is most common).

Rural tourism facilities are by nature isolated, potentially making repairs and replacement difficult and expensive. Cutting installation costs at the start of a project by not hiring specialists, decreasing the warranty period, using inferior installation equipment, and buying lower quality equipment may result in higher maintenance and repair costs in the future. In order to help ensure the sustainability of the system, specify requirements for vendor/installer training of local staff in the contract (more details are provided in the following step).

## Energy System Finance Options

In spite of the growing number of proven on-site power generation projects, many financial institutions are reluctant to lend money for these efforts as they are unfamiliar with the technologies and lack methods to assess the risks associated with such loans. However, lenders are developing more experience with small power projects and several international and government programs are promoting renewable energy finance in an effort to address this barrier.

As governments recognize the important role of tourism to national economic development, they are making institutional arrangements to support the financing of tourism projects. For example, early in 2008, the Government of El Salvador announced an agreement between the national Tourism Ministry's Salvadorian Tourism Corporation and the Agricultural Development Bank to provide financing for ecotourism. The agreement resulted in the creation of a new credit line specifically designed for ecotourism facility operators.

Energy supply system financing can be based on recovering investment costs through profits from the sale of tourism services, or in some cases, by adding a revenue stream through the sale of energy services. For example, a remote facility near a community without electricity could sell battery-charging services to individual consumers or refrigeration to local fishing businesses. Such an arrangement would have to comply with local regulations since some countries and localities have laws preventing private operators from selling electricity. Options for financing a power system project include:

- **Direct investment:** When a facility has access to sufficient cash, it can invest in its own power system project. The facility pays cash to the equipment supplier for installation and to a maintenance service provider for ongoing maintenance services throughout the life of the project.
- **Loans:** Banks or other lending institutions familiar with small power system projects lend money to the tourism facility or system supplier/local dealer. The tourism facility repays the loan either to the lending institution or supplier. In the latter case, the supplier takes on the risk for the loan. A benefit of receiving a loan for a local dealer is that they can offer service support to the unit after the sale.
- **Third-party financing:** This involves a partnership between the tourism facility owner, energy supplier, and a for-profit or non-profit financial institution. The supplier and financial institution work together, contributing their complementary experience to the project. The supplier provides technical expertise and installation and maintenance services, while the financial institution contributes financial expertise and services. This arrangement allows each to participate in projects without having to develop new areas of expertise. The financial institution provides a loan to the tourism facility based on the facility's credit history, and the supplier receives cash for installation services.
- **Fee-for-service:** In this arrangement, the system supplier is an energy service company (ESCO) that owns the system and recovers costs through payments from the tourism facility. The ESCO provides financial and installation services, and operates and maintains the system throughout the project, absorbing the project's financial risk. ESCOs can include community-owned small electricity cooperatives, rural electric cooperatives, privately- owned generation and distribution companies, and NGO-owned and operated systems.
- **Leasing:** Leasing can be an option for a system that is similar to a fee-for-service system. The facility will operate the equipment, but if there are any issues, the leasing company will often provide maintenance and repairs. The equipment remains the property of the leasing company, unless there is a lease-to-own option available.

These financing options can also be used if a facility would like to upgrade existing appliances and equipment to more efficient models. Refer to the References Section for more information on organizations working to increase access to affordable finance.

## Regulations and Permitting Issues

National and local laws and regulations affect on-site power generation projects because they govern natural resource and land use, environmental impact, and power generation. Laws may prevent certain types of activities or equipment installations, or require payment of permitting fees for installation and operation of certain types of equipment.

As a project developer, a tourism facility must work with local and regional regulators in the early stages of project development to find out which types of zoning, energy, and environmental regulations apply to the project. Communities in the area that are affected or feel affected by a power project should also be involved in the project from an early stage to address potential opposition to the project. Stakeholder acceptance, especially at the local community level, is key to a successful project. Factors that may be regulated by laws or causes for concern can include:

- Noise and air pollution from generators.
- Hazardous waste disposal for batteries and used motor oil. ▫ Soil and groundwater pollution from fuel or oil spills.
- Wildlife habitat impact, including potential disruption of bird nesting by wind systems and potential interference with stream and river habitat by hydropower systems.
- Noise and visual impact of wind systems.
- Natural resource impact of biomass-based power systems or wood-burning space heaters, cooking, and water heating.
- Potential soil erosion caused by hydropower systems. ▫ Laws against on-site power generation.

# STEP 7. MONITOR, OPERATE, AND MAINTAIN THE ENERGY SYSTEM

Once your energy supply equipment and appliances have been installed, they must be properly monitored, operated, and maintained. On-site power generation equipment and major energy-using equipment, such as heating and cooling systems, require attention on a regular basis to keep working efficiently.

To successfully operate and maintain its energy systems, every facility should have a monitoring program to track and gauge its energy performance. An effective energy monitoring program is a planned and documented system that is used to collect and analyze information on the amount of energy consumed by the property. It can help a property define normal energy use patterns, identify any shifts in consumption which could indicate the existence of operational or equipment problems, ensure the staff is complying with the energy efficiency measures adopted by the facility, and measure the results of energy conservation efforts.

In its simplest form, a monitoring program in an off-grid facility should track the consumption of all sources of energy that can be readily measured. At the end of each month, the facility should use the collected data to:

- Determine the total monthly consumption for each energy source.
- Determine total monthly guest and staff occupancy (measured in person-days).
- Use the monthly energy consumption and occupancy data to calculate monthly energy consumption indices for each energy source.
- Compare these indices with those calculated for the preceding months or for the same month of the previous year.
- Investigate any large and unexpected changes in total energy consumption or energy use indices.

At the end of each year, the facility should calculate annual consumption and consumption indices for each energy source. These annual figures can then be compared with those calculated for the preceding years to gauge the long-term trends in the facility's energy use.

It is very important to share energy monitoring information with the staff. This provides a constant reminder that energy efficiency is an important concern and gives workers feedback on the results of their efforts.

Operating a power generation system consists of daily tasks required to keep equipment in top condition. Tourism facility personnel should be designated to manage operating tasks which include ordering fuel and keeping fuel tanks full, cleaning and inspecting equipment, turning equipment on and off, keeping operation and maintenance logs, and keeping staff skills up-to-date. Costs include labor, fuel, warranty fees, service contracts, and spare parts (depending on the system and location, the facility may keep a small stock of spare parts). Operation and maintenance logs and repair manuals should be as clear and simple as possible, containing checklists, step-by-step explanations, and illustrations. Because personnel may change frequently, this information needs to be easily available and understandable so it can be understood by new staff.

Maintenance tasks include preventative and scheduled maintenance which, depending on the type of system, may involve performing engine oil changes, cleaning solar panels and wind turbine blades, and checking electrical connections and battery electrolyte levels. The on-site operator should have the skills and knowledge to perform basic preventative maintenance tasks. Other scheduled maintenance might have to be performed by qualified professionals through a maintenance service contract.

End-use equipment should be operated properly and maintained periodically to maximize energy efficiency. As some appliances age, they become less efficient. Maintaining or repairing the appliances may improve their efficiency, but when energy savings do not justify maintenance costs, it may be worth replacing the equipment. It is extremely important to monitor systems; to be done correctly, this may require the installation of special equipment (such as meters). The monitoring equipment can be added later if it is not installed initially, but including it from the beginning will help in setting up the monitoring and maintenance plan.

## Operation and Maintenance Personnel

At least one employee should have responsibility for overseeing operation and maintenance of on-site energy supply equipment, as well as facility energy use, with one or more additional employees trained as backup. It is extremely important to create an energy management team with clear roles for maintenance and accountability to keep the system operating correctly and efficiently. Once an energy management team is established, expertise can be matched to specific roles and a dedicated list of tasks/responsibilities developed.

On-site energy supply maintenance tasks include identifying equipment problems and knowing who to contact for scheduled maintenance and unscheduled repairs. Most unscheduled maintenance and repairs will need to be performed by qualified professionals. The on-site operator should be able to identify maintenance issues and shut down equipment in the event of a breakdown. In some cases, the facility manager or owner may have a particular interest in energy technologies and be qualified to operate and maintain the energy supply systems. However, in most cases, the facility will have to rely on outside expertise for equipment maintenance and repairs. The installation and maintenance contract for all on-site energy supply equipment should include training for facility staff in daily equipment operation and maintenance. Equipment suppliers and maintenance contractors should be required to provide simple, clear reference materials and easy-to-fill-in maintenance logs and other data-entry forms for local staff. Accurate record-keeping and reference materials are critical.

## CASE STUDY

### Three Camel Lodge Managing Multiple Energy Sources

The Three Camel Lodge in Mongolia's Gobi Desert provides visitors with a traditional living experience by housing them in gers, the tents used by nomadic herders in the area. The tents are hand-made using a latticed wood frame and layers of felt and canvas.

Installing the solar hot water system at the Lodge (Top); the Three Camel Lodge highlights traditional Mongolian architecture (bottom).

Most of the lodge's 48 gers rely on shared bathroom facilities, while 1 8 deluxe gers provide en-suite bathrooms but have communal showers. The lodge hosts visitors in the summer as well as during a one-week winter camel festival. The growing popularity of Three Camel Lodge has increased visitor traffic, resulting in greater energy and water demands.

The facility has a 1,900 watt solar system to power lighting in the gers and a common area, and a television and audio equipment in the common area. Two small wind turbines (Chinese models rated at 600 watts and 1,000 watts each) provide supplemental power whenever there is wind. A diesel-powered generator provides back-up power, supplies electricity to a refrigerator and freezer in the kitchen, and powers a water pump for a communal well, which supplies both the lodge and neighboring families. These families use the water at no charge for domestic purposes and livestock support.

Photo courtesy of Pam Baldinger.

Hot water for the facility is provided by a solar hot water system that uses coal as a back-up fuel. Heat for gers during the summer tourist season is provided by traditional dung-burning stoves that serve as space heaters. The dung is collected around the lodge, where herds belonging to the neighboring families graze. Because the families also use the dung for fuel, the lodge takes only what is left over after the families have met their fuel needs. During the winter festival week, the lodge uses coal for space heating, because of its superior energy content. Annual coal use is about 10 tons.

Having a wide array of energy sources offers both advantages and disadvantages. Solar and wind power have the benefit of generating electricity without consuming fuel, and produce no pollution. The solar and wind systems often complement each other, because cloudy periods tend to be windy. When there is neither wind nor sun, the diesel generator can provide power. On the other hand, having three different power sources means that staff must have the skills to operate and maintain each type of system. At the time of this writing, one of the wind turbines is broken, and awaiting repairs by a local company that has promised to have the repairs completed in time for the peak season.

Topics for maintenance training include:

- Review of equipment manuals.
- Hands-on practice with routine maintenance tasks.
- Distinguishing between problems that can be solved on-site and those that require a qualified practitioner to solve.
- Procedures for requesting equipment servicing, including lists of local qualified service providers. ▫ Emergency shut-down procedures.
- Review of safety precautions.
- Review of correct operating procedures to ensure the system will not be overloaded. ▫ Safety rules and regulations.
- Record maintenance (log sheets, etc.).

Lack of maintenance of power generation equipment can result in reduced energy output and even lead to complete failure. Studies of on-site power generation system failures show that a lack of trained on-site staff is a key cause of system breakdown. Training should also include correct system operations to ensure that the management team is looking out for improper use that could reduce performance. Improper use could include charging batteries at the wrong time, use of electricity by staff for personal purposes, and deep discharging of battery systems during peak load hours. Additionally, operation and maintenance personnel will need to be cautious of the potential for theft, which can include stolen energy generation equipment and/or stolen power. Theft of generation equipment can be an important issue in remote areas if there is no mechanism for system monitoring.

**Operation and Maintenance Considerations by Technology.**

| Applications | Self-Maintenance | Professional service |
|---|---|---|
| Solar electric systems | • Weekly system monitoring to assure proper battery charging<br>• Weekly panel cleaning depending on weather and climate<br>• Weekly wiring and electrical connections check<br>• Weekly battery maintenance: add water; clean contacts; remove corrosion | • Monthly maintenance check by qualified professional<br>• Three- to 10-year battery replacement, and 5- to 10-year inverter replacement<br>• Replacement of panels (25-year panel life) |
| Solar Heating Systems | • Daily performance monitoring, ensuring proper operation, heat delivery from collector, proper pumping and circulation of fluid<br>• Daily check of start-up and shut-off times<br>• Seasonal adjustments for heating applications—use vacation setting if system is unused for some time | • Replace antifreeze every 7-10 years depending on acidity<br>• Inspect collectors and tank every 5 years<br>• Replace pumps after 10 years |
| Wind power systems | • Daily monitoring of systems performance<br>• Weekly lubrication of moving parts, check bolts and guy wires, check wiring and electrical connections | • Initial 3-month maintenance check by qualified professional<br>• Check ups thereafter depending on manufacturer recommendations<br>• Ten-year turbine life |
| Engine-generator sets | • Weekly oil and coolant inspection; check for leaks<br>• As use dictates, change engine oil and filter<br>• Monthly maintenance check<br>• Annual maintenance check | • Based on hours of operation, periodic tune-ups and engine overhauls are scheduled routinely once operating hours established<br>• Rebuild engin |

## CASE STUDY

### Chumbe Island Total System Sustainability Planning

Chumbe Island is a pristine fossil coral island off the coast of Zanzibar, Tanzania. The Chumbe Island Coral Park has minimal impact on the environment—the lodge and seven guest bungalows have incorporated aspects of energy efficiency, renewable energy, sustainable water use, natural design, and recycling to create a tourism facility that is a highlight of sustainable architecture and operations. Each building functions as a self-sufficient unit that generates its own water and energy, with rainwater catchment and filtration, solar water heating, and photovoltaic electricity.

The orientation and design of the buildings makes maximum use of the Indian Ocean trade winds—instead of air conditioning, guests regulate the temperature of the bungalow with hand-woven mat panels. All the electric devices used on the island are energy efficient—including CFL bulbs and use of non-paraffin candles (the management has made one "compromise" by using 12V halogen spot lights in the bungalows since they have found this light more attractive than that from CFL bulbs).

The visitor center takes advantage of trade winds to cool the facility. © Craig Zendel.

Chumbe Island is powered 100 percent by solar electricity. The PV panels power lights, VHF radio communication, little cooling fans, cell phone and camera chargers, and a freezer in the restaurant kitchen. A small generator is used only during closure time for renovations and maintenance (for drilling machines). All water pumps on the island are hand pumps, for which no electricity is needed. Solar thermal systems heat water for showers. All laundry is done off the island. The island is run very frugally, but even so, the resort has added more solar panels over the years for new systems such as the freezer and the fire water pump. Each system (installed on a number of different buildings) is self-contained—including panels, a charge controller, and batteries. The systems include:

- Main building (Education Center): 14 panels, each 50W = 700 W total
- Each of the seven bungalows: two panels, each 50W = 700 W total
- To run the freezer in the kitchen: two 50W panels and four 80W panels = 420 W total
- Manager's house: two panels, each 50W = 100 W total
- Staff housing: four panels, each 50W = 200 W total
- Pump for fire equipment: one 50W panel = 50 W total

Along with two other staff, a full-time Technical Manager oversees the electricity systems on the island. They have all been trained on the job over the course of the project (the panels were installed in 1998). Outside experts occasionally provide advice for new installations.

Overall, management is pleased with the energy system. According to the hotel manager, "one big advantage of our solar-powered energy system was that Chumbe Island operations were not affected by the recent total electricity blackout in Zanzibar—and it lasted over a month!" The negative aspects of the system have been the high initial costs of the equipment (around $50,000), and the amount of training and maintenance required to run the systems. There were also some issues with corrosion of connections due to the salty sea climate. Initially, when the freezer was first introduced, it did not function well because it was opened too frequently, which raised the internal temperature. It took more power than anticipated to keep the temperature at the desired level, which drained the batteries and reduced their capacity. This problem has been abated through better management of kitchen staff.

Despite the high capital costs of renewable energy equipment, Chumbe Island's energy operating costs compare very favorably to those of three similarly sized lodges in off-grid remote areas in Zanzibar. For instance, Lodge A, which uses a diesel generator for only six hours a day, spends close to US$ 3,000/month on fuel, while Lodges B and C, running large generators for 24 hours/day, spend as much US$ 8,000 – 10,000/month on diesel (these prices do not include transport costs). The amount spent on fuel alone by Lodges B and C nearly equals the total overall average monthly running costs of Chumbe Island (approximately US$ 10,000). (It should be noted that Lodges B and C, with their greater energy-generating capacity, likely offer different amenities than Chumbe).

For cooking needs, Chumbe staff experimented with a wide range of cookers: two models of solar cookers—a solar box cooker and a parabolic solar-cooker; a low-pressure gas cooker; kerosene cookers; and traditional charcoal stoves. A combination of gas cookers and an energy-saving wood stove specially built for the staff kitchen was selected as the most appropriate and cost-effective option.

Additional highlights of the facility include rainwater catchment to use rainwater that is collected by the palm-thatched roofs and filtered through combined gravel and sand filters. This water is stored in large cisterns under the floor of each building. All the bungalows have composting toilets that recycle human and organic waste in a sustainable way and use no water for flushing. All plant and kitchen waste is composted and used in the composting toilets and as fertilizer in the grey water vegetative filtration plant beds. The surplus is taken from the island and given to local farmers in Zanzibar.

The managers work with staff and guests to ensure proper use of the facilities. Guests receive a folder in the bungalows that explains how to do everything. If they leave the lights on all night, they will discharge the solar battery and they will have to wait for the sun to recharge it. There have been no problems with the acceptance of the system by guests.

## CONCLUSION

Sustainable tourism has become an important trend in the tourism industry and will only continue to gain momentum as awareness of climate change increases. Each year, more and more tourists are taking environmental concerns into account when selecting their travel destinations and, in response, many tour operators are working with their suppliers to improve their environmental performance. Facilities that embrace sustainability from the start not only will be able to minimize operating costs, but also improve their standing and appeal in a competitive tourism market.

As a sector with substantial environmental impact, the tourism industry provides many opportunities to use and benefit from sustainable energy systems. A well-designed, managed, and operated energy system, coupled with energy-efficient devices and efficient use, can make all the difference to a rural tourism facility. To sustainably—both economically and environmentally—meet energy needs, a facility will often need to rely on on-site energy supply systems that minimize the use of transported fuel. Taking the time to plan and operate a sustainable facility will pay off in terms of reduced fuel costs, increased energy reliability, and improved staff and guest satisfaction. Energy issues should be integrated into tourism project planning from the beginning, in order to maximize potential savings and avoid costly delays that can result from inadequate preparation or unfamiliarity with technical issues.

Facility operators should work with energy experts and professionals to ensure that they are correctly planning for and installing the best energy options to meet the facility's requirements—taking into account facility type, size, location, and style. Working with energy professionals can help facility operators understand how and why to carry out specific tasks and can make the system seem less complicated.

Including staff and guests in the sustainable operations of the facility can both improve the functionality and efficiency of the energy system, and communicate the sustainability message. Sharing information and experiences will accelerate the acceptance of these practices—and give guests a unique tourist learning opportunity. Additionally, communicating energy management successes to other tourist facilities can expand the sustainable tourism market.

The steps and case studies included in this guide demonstrate that sustainable energy use at tourism facilities can be cost-effective, reliable, and environmentally sustainable for a wide range of applications.

# GLOSSARY

**Alternating Current (AC)** – An electric current flowing in both directions—most commonly from power generation plants.

**Battery** – A device that stores energy and makes it available in an electrical form.

**British thermal unit (Btu) –** A measurement for the energy content of fuels. Btu per hour (Bth/h) is a measure of the power of heating and cooling systems, sometimes abbreviated as Btu.

**Capital Cost** – The initial cost to purchase and install equipment.

**Charge Controller –** Controls the flow of current to and from the battery to protect from over charging or over discharging.

**Direct Current (DC)** – An electric current flowing in one direction—most commonly from solar panels or small generators.

**Lifetime** – The length of time a piece of equipment or system is expected to operate before failure due to age. Photovoltaic and wind equipment lifetimes are typically expressed in years; engines in hours of operation; batteries in number of charge-discharge cycles.

**End Use Equipment** – Devices that consume electricity or fuel to do useful work.

**Engine-Generator Set** – An internal combustion engine coupled to an electric generator that produces electricity. Also known as generator or diesel generator.

**Energy** – The capacity of a physical system to do work.

**Energy Efficiency** – Using less energy to provide the same level of energy service.

**Fossil Fuel** – Non-renewable fuels derived from underground hydrocarbon sources such as coal, oil, and natural gas.

**Gasifier** – A device that converts solid biomass into a gas to burn for heat or electricity.

**Grid** – The network of transmission lines, distribution lines, and transformers used in central electric power systems. Off-grid locations are not connected to a grid.

**Hybrid Energy System** – An energy system that is composed of more than one type of electricity generation type, such as a combination of a generator and a PV system

**Insolation** – The amount of sunlight falling on an area over the course of a year.

**Inverter** – A device that converts DC electricity to AC electricity.

**Kilowatt (kW)** – One thousand watts. A measure of power often used to describe the electricity-generating capacity or size of an electric generation system.

**Kilowatt Hour (kWh)** – A unit of energy representing the quantity of work performed by one kilowatt of electric power in one hour.

**Operating Cost** – The day-to-day expense of using and maintaining property, including fuel, labor, and spare parts.

**Passive Solar Heat** – Utilizing the sun to directly heat water or a space.

**Phantom Loads** – Electric power consumed by electronic appliances while they are switched off or in a standby mode.

**Photovoltaic (PV) System – A** system that produces electricity from sunlight; commonly referred to as "solar electric."

**Power Rating** – Indicates the rate at which the device converts electrical energy into another form of energy, such as light, heat, or motion.

**Renewable Energy** – Energy derived from non-fossil fuel resources; includes energy produced by photovoltaic equipment, wind turbines, water-powered turbines, and biomass systems.

**Thermal Power** – Power used for heating or cooling.

**Watt (W)** – A unit of power equal to one joule per second. The size or capacity of electric power equipment and devices is sometimes given in Watts.

**Wind Turbine** – A device that converts the energy of moving air into electricity.

# REFERENCES

## On-Site Power Generation

[1] Allderdice, A. & Rogers, J. (2000). Renewable Energy for Microenterprise. Golden: National Renewable Energy Laboratory. http://www.nrel.gov/docs/fy01osti/26188.pdf.

[2] Natural Resources Canada. RETScreen International Clean Energy Project Analysis Tools. http://www.retscreen.net

[3] USAID. Powering Health: Electrification Options for Rural Health Centers. http://pdf.usaid.gov/pdf docs/PNADJ557.pdf.

## Renewable Energy Resources

[4] 3Tier Group. FirstLook online wind and solar maps. http://firstlook.3tiergroup.com.

[5] National Renewable Energy Laboratory (NREL). Overall energy analysis, modeling, renewable energy technology information. *http://www.nrel.gov* the HOMER tool is available at: http://analysis.nrel.gov/homer

[6] United Nations Environmental Program (UNEP). SWERA wind and solar maps. http://swera.unep.net.

[7] World Bank RETooIKit for Stand-alone Projects. *http://go.*worldbank.org/ZJNBLHEO50.

## Energy Efficiency

[8] ENERGY STAR. Information on both energy efficient products and practices. www.energystar.gov for information on labeling: *http://www.* energystar.gov/index.cfm?c=partners.intl implementation.

[9] U.S. Department of Energy. A Consumer's Guide to Energy Efficiency and Renewable Energy. http://www.eere.energy.gov/consumer.

[10] Wilson A, Thorne J, Morrill J. 2003. Consumer Guide to Home Energy Savings. Washington: American Council for an Energy Efficient Economy available at: http://www.aceee.org.

## Energy and Tourism

[11] Baldinger P. 2006. Energy and Sustainable Tourism: Energy Supply and Use in Off-grid Ecotourism Facilities. Washington: U.S. Agency for International Development.

[12] Green Globe International: *The green labeling program focusing on tourism*. http://www.greenglobeint.com.

[13] Perera O, Hirsch S, Fries P. 2003. *Switched On: Renewable Energy Opportunities in the Tourism Industry*. Paris: United Nations Environmental Programme. http://www.unep.fr/pc/tourism.

[14] Renewable Energy in Tourism Initiative (RETI). *Best practices information*. http://www.renewabletourism.org.

[15] The Travel Foundation. *Training toolkit for companies.* http://www.thetravelfoundation.org.uk/training for change.asp.

[16] E+Co. *Supports small and medium enterprises that supply sustainable energy*. http://www.eandco.net.

[17] *The Renadewable Energy and Energy Efficiency Partnership* (REEEP). http://www.reeep.org.

[18] Sustainable Energy Finance Initiative (SEFI). *Energy efficiency and renewable energy finance*. http://sefi.unep.org.

# APPENDIX 1. ESTIMATING ENERGY USE

To estimate daily energy use, fill out the chart below. To obtain the information for column B, read the label on the equipment or use the estimates provided in Appendix 2. This chart should not be used in lieu of a professional audit, but can give facility managers a preliminary estimate of their facility's daily energy use and help in planning purposes.

| Description of Device | A Number of Devices | B Power[2] (measured in watts) | C A x B = total watts | D On-time[3] (hours/day) | E C x D = total wattage in hours/day (total watts x on-time) | F E/ 1000 = kWh/day |
|---|---|---|---|---|---|---|
| | | | | | | |
| | | | | | | |
| | | | | | | |
| | | | | | | |
| | | | | | | |
| | | | | | | |
| | | | | | | |
| | | | | | | |
| | | | | | | |
| | | | | | | |
| | | | | | | |
| | | | | | | |
| | | | | | | |
| | | | | | | |
| | | | | | | |
| Totals | | | | | | |

# Appendix 2. Average Power Ratings and Energy Usage for Various Devices

The charts below provide power consumption estimates for various types of electrical equipment commonly found in tourism facilities. These figures may be used to complete the table in Appendix 1, but are intended only as a guide. Actual figures will vary based on the size, age, condition, and efficiency rating of each model.

## Sample Power Ratings for Appliances with Average Efficiency.

| | **Power (watts)** |
|---|---|
| Air Conditioner—Room | 1000 |
| Air Conditioner—Central | 2000 – 5000 |
| Refrigerator | 100 – 750 |
| Freezer | 100 – 300 |
| Clothes Washer | 350 – 500 |
| Clothes Dryer | 1800 – 5000 |
| Dehumidifier | 750 |
| Desktop Computer | 150 |
| Laptop Computer | 75 |
| TV (color, 20") | 150 |
| Small Laser Printer | 300 (printing); 150 (idle) |
| Small Photocopier | 250 (copying); 60 (idle) |
| Satellite Phone | 25 |
| Cell Phone Charger | 25 |
| Fan (ceiling or floor) | 75 |
| Hairdryer | 1200 – 1800 |
| Electric Shaver | 20 |

## Lighting Power Ratings.

| | **Power (Watts) Inefficient model** | **Power (Watts) Efficient model** |
|---|---|---|
| Small bulb | 40 (incandescent) | 1 1 (CFL) |
| Medium bulb | 60 (incandescent) | 14 (CFL) |
| Large bulb | 100 (incandescent) | 25 – 30 (CFL) |

## Energy Use Comparisons between Inefficient and Efficient Models

This table compares new, standard models to new efficient models of equipment. Older models will tend to be less efficient than newer equipment. For example, a new, standard 18 cu ft refrigerator might consume 465 kWh/year while an equivalent-sized efficient model will consume 411 kWh/ year. However, a similar-sized refrigerator purchased in 1984 might consume 1,457 kWh/year. Older appliances can be a huge crain on energy.

(Note: some devices cycle during the day and are "on" continuously; others have annual use assumptions. Energy use is measured is kWh per year in this table.)

| | **Energy Use (kWh/year) Inefficient Model** | **Energy Use (kWh/year) Efficient Model** | **Use Assumptions** |
|---|---|---|---|
| Refrigerator (18 cubic feet) | 465 | 411 | On continuous cycle |
| Freezer (22 cubic feet) | 799 | 719 | On continuous cycle |
| Room air conditioner, small (10,000 BTU/hour) | 2985 | 1982 | On continuous cycle |
| Submersible water pump (11,000 liters/day, 5 m water depth) | 547 | 365 | On continuous cycle |
| Clothes Washer | 573 | 267 | 392 loads per year |
| Clothes Dryer | 950 | 898 | 416 loads per year |
| Dehumidifier | 1064 | 85 1 | 1620 hours per year |
| Desktop Computer | 187 | 76 | Computer always turned off at night |
| Desktop Computer | 742 | 90 | Computer never turned off at night |
| Laptop Computer | 38 | 15 | Turned off at night |
| TV (color, 20") | 120 | 87 | On 4 hours a day |

All data from US Environmental Protection Agency, US Department of Energy, EnerGuide Appliance Directory 2007.

# Appendix 3. Sample Calculations for a Performance-based Incentive Program

To implement this type of incentive program, the facility must first calculate its baseline energy performance by determining its average energy consumption per person per day during each of the last four quarters, and then use this baseline to calculate the quarterly energy cost savings and the amount of the awards.[4]

This type of program offers many benefits. First, it drives all staff members to work together toward a common goal. This collaboration is important as all employees—maintenance workers, housekeepers, cooks, waiters, receptionists, and office personnel—have an impact on energy consumption and can contribute something toward conservation.

Second, it can yield substantial financial benefits to employees (especially where wages are low) but places no additional financial burden on the facility. The awards are paid out only if there are real and measurable reductions in energy costs, and employees are rewarded for their efforts with part of the financial savings that would not exist without the incentive program.

And third, the evaluation process is transparent since the parameters needed to determine the awards are tangible and easily measurable, such as staff and guest occupancy, the amount of gas and other fuels consumed by the hotel, and the average market value of these energy sources.

**Background:** The facility considered in this example is a 15-room off-grid hotel, which operates year-round and employs 11 workers, three of whom live in staff accommodations built next to the hotel.

**Energy use:** The electricity used by the hotel and staff accommodations comes from a photovoltaic array which supplies ~25 percent of total electricity consumption, and a diesel generator which supplies the rest. Solar water heaters produce the hot water consumed in the kitchen and staff rooms, and propane is used by kitchen appliances and the guestroom water heaters.

## Calculating the Facility's Total Equivalent Occupancy for Each Quarter of 2007

The calculations of the facility's total equivalent occupancy are shown in the table below. These calculations are based on the following inputs and assumptions:

- Number of staff living off property = 8
- Number of staff living on property = 3 those that are directly purchased by the facility, such as the fuel and gas consumed by generators and appliances. Therefore, the electricity generated by the facility's photovoltaic panels or generators, as well as the thermal energy produced by its solar water heaters are excluded from these calculations.
- The calculation of the "equivalent day staff occupancy" (column B) is based on the standard assumption that the impact of day workers on the facility's energy consumption is approximately half of that of an in-house worker or guest.
- The number of guest-nights shown in column D is obtained from the property's occupancy log.

| A | B | C | D | E (= B + C + D) |
|---|---|---|---|---|
| **Period** | **Equivalent day staff occupancy (capita-days)** | **In-house staff occupancy (capita-days)** | **Guest occupancy (capita-days)** | **Total equivalent occupancy (capita-days)** |
| 1st quarter 07 | 90 days x 8 workers x 0.5 = 360 | 90 days x 3 workers = 270 | 1,201 | 1,831 |
| 2nd quarter 07 | 91 days x 8 workers x 0.5 = 364 | 91 days x 3 workers = 273 | 892 | 1,529 |
| 3rd quarter 07 | 92 days x 8 workers x 0.5 = 368 | 92 days x 3 workers = 276 | 672 | 1,316 |
| 4th quarter 07 | 92 days x 8 workers x 0.5 = 368 | 92 days x 3 workers = 276 | 1, 110 | 1,754 |

## Calculating the Facility's Baseline Energy Performance for Each Quarter of 2007.

| A | B | C | D (=C/B) | E | F (= E/B) |
|---|---|---|---|---|---|
| **Period** | **Total equivalent occupancy (capita-days)** | **Propane consumption (L)** | **Propane cons. index (L/cap-day)** | **Diesel consumption (L)** | **Diesel cons. index (L/cap-day)** |
| 1st quarter 2007 | 1,831 | 488 | 0.267 | 2,353 | 1.29 |
| 2nd quarter 2007 | 1,529 | 468 | 0.306 | 2,083 | 1.36 |
| 3rd quarter 2007 | 1,316 | 442 | 0.336 | 1,958 | 1.49 |
| 4th quarter 2007 | 1,754 | 410 | 0.234 | 2,375 | 1.35 |

The facility's baseline energy performance for each quarter of 2007 is calculated in the table below. These calculations are based on the following inputs:

- The property's total equivalent occupancy figures are as shown in the preceding table.

## The propane and diesel consumption figures are obtained from the property's daily fuel consumption logs

## Calculating the Energy Savings Achieved in the 1st Quarter of 2008

In January 2008, the hotel launched a performance-based incentive program that was designed to share with the staff 25 percent of the energy costs savings achieved during each quarter of 2008.

| Period | Staff and guest occupancy (capita-days) | Total equivalent occupancy (capita-days) | Propane cons. (L) | Propane cons. index (L/cap-day) | Diesel cons. (L) | Diesel cons. index (L/cap-day) |
|---|---|---|---|---|---|---|
| 1st quarter 2008 | Day staff = 720 x 0.5<br>In-house staff = 270<br>Guests = 1,251 | 1,881 | 391 | 0 208 | 1,727 | 0.918 |
| 1st quarter baseline | | | | 0 267 | | 1.29 |
| Per capita energy savings | | | | 0 059 | | 0.372 |

## Calculating the Awards Resulting from the Energy Cost Savings Achieved in the 1st Quarter of 2008

| | |
|---|---|
| Propane savings in 1 st quarter | = (propane savings/capita) x (total occupancy) = (0.059 L/cap-day) x (1,881 cap-day)<br>= 111 L total savings |
| Diesel savings in 1st quarter | = (diesel savings/capita) x (total occupancy) = (0.372 L/cap-day) x (1,881 cap-day)<br>= 700 L total savings |
| Total 1st quarter financial savings | = (diesel savings/capita) x (total occupancy) = (0.372 L/cap-day) x (1,881 cap-day)<br>= 700 L total savings |
| Award amount | = 25 percent x (US$ 1 .015) = US$ 254 |
| Net savings achieved by hotel | = US$ 1 .015 - US$ 254 = US$ 761 |

According to the terms of the incentive program, 25 percent of the net cost savings in propane and diesel are shared with the staff.

The average cost of the propane and diesel purchased during the 1st quarter of 2008 is US$ 0.95 /L and US$ 1.30 /L, respectively.

## ACKNOWLEDGMENTS

We would like to express our deep appreciation and gratitude to a number of individuals who shared their knowledge and experience in the production of this publication, including the staff and managers at the facilities highlighted as case studies: Ngorongoro Serena Lodge, Tiamo Resorts, Posadas Amazonas, the Three Camel Lodge, and the Chumbe Island Coral Park.

The Global Sustainable Tourism Alliance (GSTA) Management Partners—Academy for Educational Development (AED), The George Washington University, Solimar International, and the Nature Conservancy—provided helpful feedback. Roberta Hilbruner, GSTA CTO for Development at USAID offered numerous reviews, suggestions, and outreach with those working in the field.

Walt Ratterman with SunEnergy Power International, Costas Christ, Peter Lowenthal, Paul Gilman and Patricio Gonzalez Morel provided invaluable technical expertise. We would also like to acknowledge the significant insights and contributions shared by other experts and stakeholders.

For USAID Missions and other U.S. Government programs interested in sustainable energy and tourism, the USAID Energy Team in the Bureau for Economic Growth, Agriculture and Trade, Office of Infrastructure and Engineering is available to provide additional assistance. Contact information is provided below.

**Pamela Baldinger**

Energy Advisor
Office of Infrastructure and Engineering,
EGAT Bureau US Agency for International Development
1300 Pennsylvania Ave, NW
Washington, DC 20523
pbaldinger@usaid.gov
US Agency for International Development
1300 Pennsylvania Avenue,
NW Washington, DC 20523
www.usaid.gov

## End Notes

1 For more information on sustainable tourism, see the World Tourism Organization at: http://www.unwto.org/sdt/

2 To find wattage information for a given device, check the user's manual. It may also be written on the back of the device.

3 The typical daily "on-time" for a standard refrigerator compressor is five to six hrs/day.

4 In the case on an off-grid hotel, the energy sources that are used to calculate this baseline performance include all

# INDEX

## D



## E

## U



## V



## W



## Z